INTRODUCTORY
GEOMETRY
A BRIEF COURSE WITH REASONING SKILLS

MW00444047

INTRODUCTORY GEOMETRY

A BRIEF COURSE WITH REASONING SKILLS

ALAN WISE
University of San Diego

Saunders College Publishing
Harcourt Brace Jovanovich College Publishers
Fort Worth   Philadelphia   San Diego
New York   Orlando   Austin   San Antonio
Toronto   Montreal   London   Sydney   Tokyo

Copyright © 1990 by Harcourt Brace Jovanovich, Inc.

All rights reserved. No part of this publication may be reproduced or transmitted in any form or by any means, electronic or mechanical, including photocopy, recording, or any information storage and retrieval system, without permission in writing from the publisher.

Although for mechanical reasons all pages of this publication are perforated, only those pages imprinted with an HBJ copyright notice are intended for removal.

Requests for permission to make copies of any part of the work should be mailed to: Permissions, Harcourt Brace Jovanovich, Publishers, Orlando, Florida 32887.

ISBN: 0-15-546508-2
Printed in the United States of America

PREFACE

Introductory Geometry: A Brief Course with Reasoning Skills is a guided worktext for college developmental mathematics. The clear, informal, and nonthreatening style enables students at various reading levels to master fundamental skills. *Introductory Geometry* is for:

- an introductory geometry course.
- any mathematics course that includes some coverage of basic geometry.

In each section, students are taught fundamental skills through well-structured lessons covering specific learning objectives; solved step-by-step examples; individual sets of drill exercises keyed to the text; paired and graded practice exercises; and cumulative review exercises for skill maintenance.

The guided worktext approach allows this book to be used effectively in traditional lecture courses, learning laboratories, math labs, self-study programs, and correspondence courses.

The *Instructor's Manual with Tests* contains a test package and answers to the even numbered-exercises in the book.

Preparing a book for publication requires the effort and skill of many people in addition to the author. I am grateful to the following people for their many hours spent reading manuscript and for their valuable suggestions for its improvement: Helen Joan Dykes, Edison Community College, and Miriam Keesey, San Diego State University. I also thank Kate Pawlik who worked each exercise and problem.

Alan Wise

CONTENTS

CHAPTER 1

REASONING SKILLS

To solve the problems in this chapter you will

- Use an elimination process.
- Use inductive reasoning.
- Use deductive reasoning.
- Use indirect deductive reasoning.

1.1   USE AN ELIMINATION PROCESS

The most commonly used problem-solving skill by adults is elimination. When a good consumer wants to purchase a particular product he or she first gathers information and then uses an elimination process, based on personal preferences, to make a final decision.

Example   Carol wants to buy a used car. Her newspaper lists the following possibilities:

1982 Ford 90,000 mi $1200	**1980 Chevrolet** 60,000 mi $800
1985 Mercury 30,000 mi $2750	**1981 Chrysler** 70,000 mi $900

Which cars, if any, should Carol consider if each of her following personal preferences must first be satisfied?

- The price must be $1500 or less.

- The mileage must be less than 75,000 miles.

- Only 1981 or newer models will be considered.

Solution   Eliminate the Ford because it has more than 75,000 miles.
Eliminate the Chevrolet because it is older than 1981.
Eliminate the Mercury because it costs more than $1500.

The only car of the four listed that Carol should consider, based on her three personal preferences, is the 1981 Chrysler.   ■

Check: 1981 ⟵—— 1981 or newer
                70,000 mi ⟵—— 75,000 mi or less
                    $900 ⟵—— not more than $1500

1.1 PRACTICE

Solve each problem using an elimination process.

1. Harold wants to make a one hour doctor's appointment. The available appointment times with the doctor are listed at the right. The following are other appointments that Harold must keep today.
   - Staff meeting 8:30–9:00
   - Board meeting 10:30–11:45
   - Lunch with the boss 12:00–1:30
   - Sales conference 4:00–5:00

   If it takes 30 minutes travel time each way between the doctor's office and Harold's office, which appointment time, if any, should Harold take?

Available Appointment Times
9:45
10:45
11:45
1:15
2:15
4:15

For each problem 2–5, use the figure or table at the right to aid in the elimination process.

2. Angela, Bailey, Corinna, Dwayne, and Ernestine work together. They sit at a round table for their business meetings. No two people sit next to each other if the first letters of their names are next to each other in the alphabet. Ernestine's brother sits on her right. If Bailey and Dwayne are the only males, where is everyone sitting in relation to Corinna?*

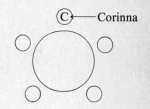

3. The three teams in the Western Volleyball League had a season in which each team played the other two both at home and away. The season's schedule is shown in the following table. Only one game was played on a given day.*

Team	Won	Lost
Spikers		
Setters		
Servers		

Day	1	2	3	4	5	6
Opponents	Spikers Setters	Spikers Servers	Setters Servers	Spikers Servers	Setters Servers	Spikers Setters

- The Spikers never defeated the Setters.
- Even though the Servers lost two games, they never lost a game at home.

What was the win-loss record of each of the three teams?

4. The greatest number of digits to which π (pi) has been memorized and recited was by Creighton Carvello of Cleveland, England, on June 27, 1980. Use the clues below to identify the correct number of digits to which π was memorized and recited by Creighton Carvello. Choose your answer from the list at right.
   - The ones digit is the largest digit.
   - The ten-thousands digit is *even* (0, 2, 4, 6, or 8).
   - The first digit is less than the last digit.
   - The thousands digit is the smallest digit.
   - The sum of the first and last digit is 5.

21,013
20,012
20,413
40,015
30,014
20,013

* Source: *Mathematics Teacher*, January 1986, (worksheet 2). Reprinted by permission.

Copyright © 1990 by Harcourt Brace Jovanovich, Inc. All rights reserved.

1.2 USE INDUCTIVE REASONING

To reason inductively, you draw conclusions based on your observations. In inductive reasoning, you look at a few of the items in a group of items and then draw a conclusion about all the items in the group. That is, you use a few specific examples to help you come up with a generalization.

Example 1   What is the next number (represented by the question mark "?") in the sequence below?

$$2, \ 3, \ 5, \ 8, \ 13, \ ?$$

Solution   To solve a problem such as this, you try to find a pattern in the sequence of numbers. First note that the sum of the first two numbers in the sequence, 2 and 3, is 5. Then note that the sum of 3 and 5 is 8. And then note that the sum of 5 and 8 is 13. Using the pattern of adding the previous two numbers together to get the next number, you should now conclude that the next number in the given sequence must be 21 because

$$8 + 13 = \mathbf{21} \quad \blacksquare$$

Example 2   What is the next number in the sequence below?

$$1, \ 6, \ 4, \ 9, \ 7$$

Solution   To solve a problem such as this, you must once again find a pattern in the sequence of numbers. However, this time the pattern is a little more complicated. First note that to get from 1 to 6, you add 5. Then to get from 6 to 4, you subtract 2. And then to get from 4 to 9, you again add 5. And then to get from 9 to 7, you again subtract 2. Using the pattern of add 5, subtract 2, you should now conclude that the next number in the sequence is 12 because

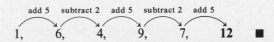

The objects in a sequence do not have to be numbers. For example, the next example deals with a sequence of *triangles*.

Example 3   What is the next triangle in the sequence below?

Solution   To solve a problem such as this, you must find a pattern in the given sequence of triangles. First note that to get from the first triangle in the sequence to the second, you rotate the first triangle counterclockwise one-quarter turn. Then note that to get from the second triangle to the third triangle in the sequence, you once again rotate the second triangle one-quarter turn counterclockwise. And then note that to get from the third triangle to the fourth triangle, or from the fourth triangle to the fifth, you continue to rotate the triangles one-quarter turn counterclockwise. Using the pattern of rotating the previous triangle one-quarter turn counterclockwise to get the next triangle, you should conclude that the next triangle in the sequence is

1.2 PRACTICE

Find the next member of each sequence below using inductive reasoning.

1. 1, 2, 3, 4, 5

2. 0, 1, 3, 6, 10

3. 0, 2, 4, 6, 8

4. 1, 2, 4, 8, 16

5. 0, 2, 4, 6, 12

6. 1, 2, 4, 8, 10

7. 9, 7, 8, 6, 3

8. 1, 4, 2, 8, 4

9. Fibonacci Sequence
1, 1, 2, 3, 5,

10. Lucus Sequence
1, 3, 4, 7, 11

11. , , , ,

12. O, T, T, F, F, S, S

13. Use the diagram below to answer the question that follows.

Which figure should come next in this sequence?

14. What is the missing figure in this sequence?

Copyright © 1990 by Harcourt Brace Jovanovich, Inc. All rights reserved.

1.3   USE DEDUCTIVE REASONING

To reason deductively, you use known relationships to deduce (find) new relationships.

Example 1   Given the four basic relationships below:

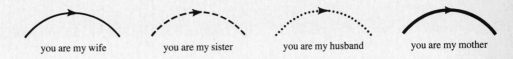

you are my wife   you are my sister   you are my husband   you are my mother

What could *E* say to *G* in the following arrow diagram?*

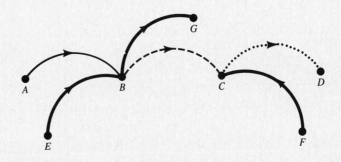

known relationships (see arrow diagram)

Solution   If $\overline{B}$ is the mother of *E*, and $\overline{G}$ is the mother of *B*, then
G must be the grandmother of *E*. ⟵—— new relationship
Therefore *E* could say to *G*: You are my grandmother.

Note: In deductive reasoning, statements of the "if-then" type are often used to reach new conclusions. (See the above Solution.) The "if" part of an "if-then" statement is called the **hypothesis** (hy-poth-eh-sis) and the "then" part is called the **conclusion.**

* Source: *Mathematics Teacher*, March 1985. Reprinted by permission.

1.3 PRACTICE

Answer each question using deductive reasoning, the four basic relationships, and the arrow diagram above.

1. What could *A* say to *C*?

2. What could *G* say to *C*?

3. What could *F* say to *B*?

4. What could *E* say to *F*?

5. What could *F* say to *E*?

6. What could *A* say to *F*?

7. What could *G* say to *A*?

8. What could *D* say to *G*?

9. What could *F* say to *D*?

10. What could *E* say to *D*?

11. What could *F* say to *G*?

12. What could *A* say to *D*?

HISTORICAL NOTE

The introduction of proof by deductive reasoning by the Greek mathematician Thales of Miletus (640–546 B.C.) was one of the most important steps in the development of human thought processes.

Copyright © 1990 by Harcourt Brace Jovanovich, Inc. All rights reserved.

USE INDIRECT DEDUCTIVE REASONING

In the previous problem solving section, you solved each problem using direct deductive reasoning. That is, you went directly from one given point in the diagram to another given point while keeping track of the given basic relationships. To solve a problem using deductive reasoning when the proposed solutions cannot be obtained easily with a direct approach, you should try to eliminate some (or all) of them using indirect deductive reasoning.

> To eliminate a proposed solution of a given problem using **indirect deductive reasoning,** you
> 1. Assume the proposed solution to be a true solution of the problem.
> 2. Show that the assumption in Step 1 leads to a contradiction (an impossible situation) using information and facts from the given problem.
> 3. Eliminate the proposed solution because assuming it to be true in Step 1 has led to a contradiction in Step 2. That is, if assuming a proposed solution is true leads to an impossible situation, then the proposed solution must really be false and as such it must be rejected.

Example   Three women each purchased a dress for a wedding. Bobbi spent $30 more than Jerry. Jerry spent $40 less than Gail. Gail spent $20 more than Bobbi. Could Bobbi have spent $80 on her dress?

Solution   Since the solution to this problem is not obvious, you might try to solve it using indirect reasoning as follows:

1. Make assumption   Assume Bobbi did spend $80 on her dress.

2. Show contradiction   • Since Bobbi spent $30 more than Jerry, Jerry must have spent $50 on her dress.

• Since Jerry spent $40 less than Gail, Gail must have spent $90 on her dress.

• Since Gail spent $20 more than Bobbi, Bobbi must have spent $70 on her dress. This is a contradiction because Bobbi spent $80 on her dress by the assumption in Step 1.

3. Reject false assumption   Because assuming Bobbi spent $80 on her dress has led to the contradiction that Bobbi spent $70 on her dress, the assumption in Step 1 must be false.

4. Interpret   Therefore, Bobbi could *not* have spent $80 on her dress.   ■

1.4 PRACTICE

Solve each problem using indirect reasoning.

To answer questions 1 and 2, use the facts from the previous Example.

1. Did Jerry spend $80 on her dress?   **2.** Did Gail spend $80 on her dress?

3. The boxes at the right are marked "R" for red marbles, "B" for blue marbles, and "RB" for a mixture of red marbles and blue marbles. If none of the boxes is labeled correctly and a red marble has just been drawn from the box marked "RB," then what is in each box? (*Hint:* Use indirect reasoning to help determine what is in the box marked "RB." Then use a direct elimination process to determine what is in the box marked "B" and "R," respectively.)

R	B	RB

4. A gumball machine contains only black gumballs and red gumballs. What is the smallest number of pennies that must be spent to be sure of getting at least two gumballs of the same color? (*Hint:* Use indirect reasoning starting with the second gumball purchased. That is, assume that each gumball after the first one does not match any of the others already purchased, and then look for a contradiction.)

5. If only one of the following statements is true, which of the four people is guilty?
 Myrle: "Evelyn did it."
Evelyn: "Diane did it."
  Alan: "I didn't do it."
 Diane: "Evelyn lied when she said I did it."
(*Hint:* Assume each statement is true (one at a time), and then look for contradictions.)

6. At a convention of liars and truth-tellers, a new friend tells you that the girl in the red dress just told him that she was a liar. Is your friend a liar or truth-teller? (*Hint:* Use indirect reasoning by first assuming the girl in red is a truth-teller, and then assume she is a liar.)

7. A rich woman's money was missing. The thief was either the butler, the maid, or the cook. During the investigation, the suspects made the following statements:
Butler: "The maid stole the money."
 Maid: "That is true!"
 Cook: "I did not steal the money."
If at least one of them lied and at least one of them told the truth, then who stole the money?

8. Three joggers, Robert, Michael, and Bill are jogging toward the country club. Robert always tells the truth. Michael sometimes tells the truth, whereas Bill never does.

Determine the names of each runner and explain how you know. (*Hint:* First, determine which one is Robert.)

Copyright © 1990 by Harcourt Brace Jovanovich, Inc. All rights reserved.

9. Four animals are kept in a barn at night. Each animal is kept in a separate stall. The animals must be put into the barn according to the following rules.
   1. If the horse is in the first stall, the cow cannot be in the second stall.
   2. The cow must be in a stall next to the sheep.
   3. The goat must always be put in the fourth stall.

   On one particular evening, the horse is in the first stall. Based on this information, which of the following conclusions is valid?
   A. The goat is in the stall next to the horse.
   B. The horse is in the stall next to the cow.
   C. The sheep is in the second stall.
   D. The cow is in the fourth stall.

10. There are 25 students in a woodworking class. During the class, 13 of the students built a bookshelf, and 11 students built a birdhouse. Five students built both a bookshelf and a birdhouse. How many students in the class did not make either a bookshelf or a birdhouse?

Copyright © 1990 by Harcourt Brace Jovanovich, Inc. All rights reserved.

CHAPTER 2

LINES, ANGLES, AND TRIANGLES

In this chapter you will

- Measure line segments and angles.
- Identify special angles and triangles.

| 2.1 | Measure Line Segments and Angles |

A **point** is an exact location in space. Technically, a point has no shape or size. However, a simple dot can be used to represent a point.

point

A straight **line** is an **infinite** (uncountable) collection of points that has no beginning and no ending. A line is a one-dimensional figure that extends infinitely far in two opposite directions. For example, the line below extends infinitely far to the left and to the right.

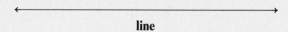

line

Note: The arrowheads show that the line continues on forever in both directions.

A **line segment** is a complete piece of a line that has a definite beginning and ending point.

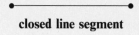

line segment

Note: No arrowheads show that the line segment has a definite beginning and ending.

A **closed line segment** contains both its end points. To indicate that a line segment contains both its end points, you draw a solid circle at each end of the line segment.

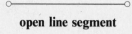

closed line segment

An **open line segment** does not contain either one of its end points. To indicate that a line segment does not contain either one of its end points, you draw an open circle at each end of the line segment.

open line segment

A **half-closed line segment** (or **half-open line segment**) contains one end point, but not the other.

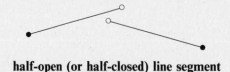

half-open (or half-closed) line segment

A **ray** (or **half-line**) is a portion of a line that has a beginning but no ending. That is, a ray extends infinitely far in only one direction.

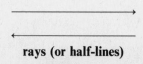

rays (or half-lines)

A **closed ray** (or **closed half-line**) contains its beginning point.

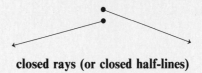

closed rays (or closed half-lines)

An **open ray** (or **open half-line**) does not contain its beginning point.

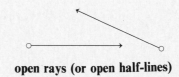

open rays (or open half-lines)

In geometry, straight lines and segments are usually denoted by letters. For example, the following line

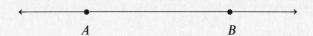

is denoted by

which is read as "line *AB*."

And the following line segment *AB*,

is denoted by

$$\overline{AB} \quad \text{or simply} \quad AB$$

which is read as "line segment *AB*," or simply "segment *AB*."

A *ruler*, such as the one shown below, can be used to measure the length of a line segment in inches (in.), centimeters (cm), or millimeters (mm).

inch ruler

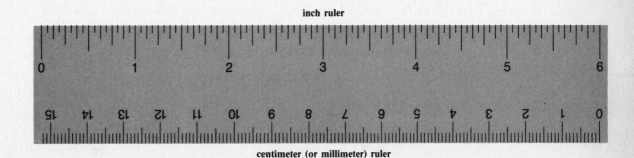

centimeter (or millimeter) ruler

When a measurement is made using a measurement tool, such as a ruler, the resulting measure is only an approximation of the true measure. In general the accuracy of a measurement will depend on the needs of the person making the measurement land the accuracy of the measurement tool being used.

Agreement: In this text it will be understood that when an equality symbol = is used to indicate a measurement, such as

$$XY = 3 \text{ in.} \quad \text{ or } \quad XY = 7 \text{ cm}$$

the indicated equality is only an approximation of the true measure.

Example 1 For the following line segment XY,

$$X \text{———————————————} Y$$

use a ruler to measure the length of XY to the nearest:

a. inch **b.** half-inch **c.** quarter-inch **d.** eighth-inch

e. sixteenth-inch **f.** centimeter **g.** millimeter **h.** tenth-centimeter

Solution Using a ruler:

a. $XY = 3$ in., to the nearest inch.

b. $XY = 2\frac{1}{2}$ in., to the nearest half-inch.

c. $XY = 2\frac{3}{4}$ in., to the nearest quarter-inch.

d. $XY = 2\frac{5}{8}$ in. or $2\frac{6}{8}$ in., to the nearest eighth-inch.

e. $XY = 2\frac{11}{16}$ in., to the nearest sixteenth-inch.

f. $XY = 7$ cm, to the nearest centimeter.

g. $XY = 68$ mm, to the nearest millimeter.

h. $XY = 6.8$ cm, to the nearest tenth-centimeter.

Verify each measurement **a** through **h** using a ruler. ■

Note: All of the previous measurements are only approximations of the true length of line segment XY. In general, it is impossible to make an exact measurement using a measuring tool such as a ruler.

When dealing with length measures, it is often necessary to convert from U.S. to metric measures, or vice versa, from metric to U.S. measures.

To convert between U.S. and metric length measures, you multiply by the appropriate conversion factor, as shown below.

U.S. to Metric Conversion Factors		
From	**To**	**Multiply By**
inches (in.)	millimeters (mm)	25.4
inches	centimeters (cm)	2.54
feet (ft)	meters (m)	0.3048
yards (yd)	meters	0.9144
miles (mi)	kilometers (km)	1.609

Metric to U.S. Conversion Factors		
From	**To**	**Multiply By**
millimeters (mm)	inches (in.)	0.03937
centimeters (cm)	inches	0.3937
meters (m)	feet (ft)	3.280
meters	yards (yd)	1.094
kilometers (km)	miles (mi)	0.6214

Example 2   The average National Basketball Association (NBA) player is 6 feet 7 inches tall. How tall is this to the nearest whole centimeter?

Solution   The question in this problem asks you to convert from 6 ft 7 in. to the nearest number of whole centimeters. To rename 6 ft 7 in. in terms of centimeters, you first rename 6 ft 7 in. as inches:

$$6 \text{ ft } 7 \text{ in.} = 6 \text{ ft } + 7 \text{ in.}$$
$$= 72 \text{ in.} + 7 \text{ in.}$$
$$= 79 \text{ in.}$$

Then rename 79 in. in terms of centimeters by multiplying by the conversion factor from the table on page 20 that relates inches and centimeters:

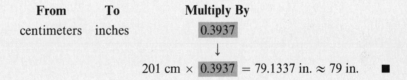

From	To	Multiply By
inches	centimeters	2.54

$$79 \text{ in.} \times 2.54 = 200.66 \text{ cm}$$

And then round 200.66 cm to the nearest whole centimeter:

$$200.66 \text{ cm} \approx 201 \text{ cm}$$

The average height of an NBA basketball player is 201 cm, to the nearest whole centimeter.

Check: Is 201 cm about 6 ft 7 in. (79 in.)? Yes:

From	To	Multiply By
centimeters	inches	0.3937

$$201 \text{ cm} \times 0.3937 = 79.1337 \text{ in.} \approx 79 \text{ in.} \quad ■$$

A **2-dimensional figure** is formed by lines that are straight, curved, and/or broken. A **simple closed figure** is a two-dimensional figure that does not cross over itself and in which you cannot get from the **inside** to the **outside** without crossing over a point **on** the figure itself.

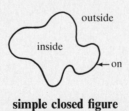

simple closed figure

A **simple open figure** is a two-dimensional figure that does not cross over itself and is not a simple closed figure. A simple open figure does not have an inside or an outside.

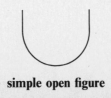

simple open figure

An **angle** is an open figure formed by two different straight line segments (or rays) that share a common end point. The common end point is called the **vertex** of the angle. The two different line segments (or rays) are called the **sides** of the angle.

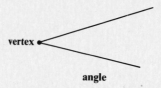

In geometry, angles are usually denoted by letters. For example, the following angle

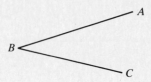

is denoted by

$$\angle ABC, \quad \text{or } \angle CBA, \quad \text{or simply } \angle B$$

which are read as "angle *ABC*," "angle *CBA*," "angle *B*," respectively.

Note: When three letters are used to denote an angle, the middle letter represents the vertex of the designated angle.

A *protractor* can be used to measure an angle in **degrees** (°). For example, the following protractor is shown measuring an angle of 1° (1 degree):

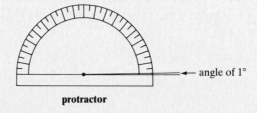

protractor

When measuring an angle, an *arc* is often used to denote the angle being measured. For example, the arc in the following angle

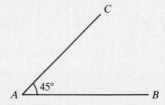

indicates that the measure of angle *BAC* is 45 degrees, which is denoted by

$$m(\angle BAC) = 45° \text{ or simply } \angle A = 45°$$

Note: Read $m(\angle BAC) = 45°$ as "the measure of angle *BAC* is 45 degrees."

Verify that $\angle A = 45°$ using a protractor.

Example 3 In the following figure,

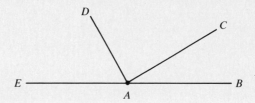

Use a protractor to measure:
a. $\angle BAC$ **b.** $\angle BAD$ **c.** $\angle BAE$
d. $\angle CAD$ **e.** $\angle CAE$ **f.** $\angle DAE$

Solution Using a protractor:
a. $\angle BAC = 30°$
b. $\angle BAD = 120°$
c. $\angle BAE = 180°$
d. $\angle CAD = 90°$
e. $\angle CAE = 150°$
f. $\angle DAE = 60°$

Verify measurements **a** through **f** using a protractor. ■

CAUTION When you are given just one angle, such as

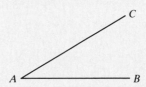

you can name the angle as $\angle BAC$, $\angle CAB$, or simply $\angle A$. However, the simplified notation $\angle A$ cannot be used when there is more than one angle with the common vertex A. For example, the notation $\angle A$ makes no sense in Example 3 because there are at least six different angles with the vertex A.

Sometimes angles are measured more accurately by dividing the degree into **minutes:**

$$\textbf{1 degree} = \textbf{60 minutes (or 60')}$$

And sometimes angles are measured even more accurately by dividing the minute into **seconds:**

$$\textbf{1 minute} = \textbf{60 seconds (or 60'')}$$
$$\textbf{1 degree} = \textbf{3600 seconds (or 3600'')}$$

For example, an angle of $28°15'$ has been measured to the nearest minute, while an angle of $62°45'36''$ has been measured to the nearest second. When performing calculations with angles that have been measured to the nearest minute or second, it is often convenient to rename the angle in decimal form, as an equal amount of degrees, before doing the arithmetic.

Example 4 Rename $28°15'$ in decimal form, as an equal amount of degrees.

Solution
$$28°15' = 28° + 15' \qquad \text{Write as a sum.}$$
$$= 28° + \frac{15°}{60} \qquad \text{Divide by 60.}$$
$$= 28° + 0.25° \qquad \text{Do the arithmetic.}$$
$$= 28.25° \longleftarrow \text{decimal form}$$

And so, in decimal form

$$28°15' = 28.25° \quad \blacksquare$$

Note: To reverse the process shown in Example 4, you multiply by 60 instead of dividing by 60. For example:

$$
\begin{aligned}
28.25° &= 28° + 0.25° &&\text{Write as a sum.}\\
&= 28° + \mathbf{60}(0.25)' &&\text{Multiply by 60.}\\
&= 28° + 15' &&\text{Do the arithmetic.}\\
&= 28°15' &&\text{Write in standard form.}
\end{aligned}
$$

Example 5   Rename $62°45'36''$ in decimal form, as an equal amount of degrees.

Solution
$$
\begin{aligned}
62°45'36'' &= 62° + 45' + 36'' &&\text{Write as a sum.}\\
&= 62° + \frac{45°}{\mathbf{60}} + \frac{36°}{\mathbf{3600}} &&\begin{aligned}&\text{Divide minutes by 60.}\\&\text{Divide seconds by 3600.}\end{aligned}\\
&= 62° + 0.75° + 0.01° &&\text{Do the arithmetic.}\\
&= 62.76° \longleftarrow \text{decimal form}
\end{aligned}
$$

And so, in decimal form

$$62°45'36'' = 62.76° \quad \blacksquare$$

Note: To reverse the process shown in Example 5, you multiply instead of divide. For example:

$$
\begin{aligned}
62.76° &= 62° + 0.76° &&\text{Write as a sum.}\\
&= 62° + \mathbf{60}(0.76)' &&\text{Multiply by 60.}\\
&= 62° + 45.6' &&\text{Do the arithmetic.}\\
&= 62° + 45' + 0.6' &&\text{Write as a sum.}\\
&= 62° + 45' + \mathbf{60}(0.6)'' &&\text{Multiply by 60 again.}\\
&= 62° + 45' + 36'' &&\text{Do the arithmetic.}\\
&= 62°45'36'' &&\text{Write in standard form.}
\end{aligned}
$$

2.2   IDENTIFY ANGLES AND TRIANGLES

A **right angle** (square corner) measures exactly 90°.

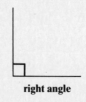

right angle

An **acute angle** measures more than 0° but less than 90°.

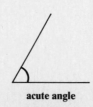

acute angle

A **straight angle** (straight line segment) measures exactly 180°.

An **obtuse angle** measures more than 90° but less than 180°.

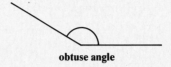

Two angles are **complementary angles** if the sum of their measures is 90°.

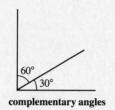

Two angles are **supplementary angles** if the sum of their measures is 180°.

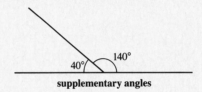

A closed figure with sides that are all straight line segments, such as the one below, is called a **polygon.**

When a line segment is part of a closed figure, such as a polygon, it is called a **side**. A **triangle** is a polygon with three sides.

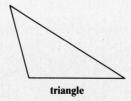

Note: Every triangle has three sides,

three vertices,

and three angles:

In geometry, triangles are usually denoted by capital letters. For example, the following triangle

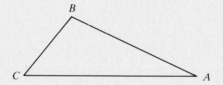

is denoted by

$$\triangle ABC$$

which is read as "triangle *ABC*."

Note: In $\triangle ABC$:

- The three sides are *AB*, *BC*, and *CA*.

- The three vertices are the points *A*, *B*, and *C*.

- The three angles are $\angle A$, $\angle B$, and $\angle C$.

For convenience, each side of a triangle is often labeled with the lower case letter of the angle opposite it.

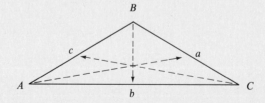

In a triangle, each angle has one **opposite side** and two **adjacent sides.** For example, in $\triangle ABC$ above:

- The side opposite $\angle A$ is *a*.

- The sides adjacent to $\angle A$ are *b* and *c*.

In $\triangle ABC$, which side is opposite $\angle B$ and which sides are adjacent to $\angle C$?

A geometric **plane** is a flat surface with no thickness whose length and width extend infinitely in all directions, without boundaries. Because a plane has the same thickness as a point (an undefined, unmeasurable thickness), you cannot see a plane. However, for convenience a chalkboard or a piece of paper, such as the one you are now reading, are often used to represent a **limited plane** (a piece of a plane).

In a geometric plane, figures that are exactly the same shape and size are called **congruent figures.** The concept of congruent figures plays such a major role in the study of geometry that a special **congruence symbol** $\cong$ is used to represent the words "is congruent to." In particular, *congruent lines* and *congruent angles* are often used to construct congruent figures. **Congruent lines** are line segments that have exactly the same length.

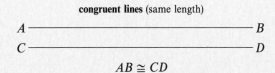

congruent lines (same length)

$AB \cong CD$

Note: Read $AB \cong CD$ as "segment AB is congruent to segment CD."

Congruent angles are angles that have the exact same degree measurement.

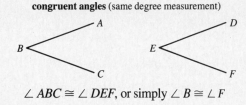

congruent angles (same degree measurement)

$\angle ABC \cong \angle DEF$, or simply $\angle B \cong \angle F$

Note: Read $\angle B \cong \angle E$ as "angle B is congruent to angle D."

One way to classify triangles is by the number of sides that are congruent. For example, an **equilateral triangle** is a triangle with three sides that are congruent to each other.

equilateral triangle

Note: In an equilateral triangle, the three angles are also congruent.

An **isosceles triangle** is a triangle with two sides that are congruent.

isosceles triangle

Note: In an isosceles triangle, the **base angles** opposite the congruent sides are also congruent. Also, every equilateral triangle is an isosceles triangle, but not every isosceles triangle is an equilateral triangle.

Copyright © 1990 by Harcourt Brace Jovanovich, Inc. All rights reserved.

A **scalene triangle** is a triangle with no two sides that are congruent.

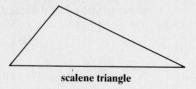

scalene triangle

Note: In a scalene triangle, no two angles are congruent either.

Another way to classify triangles is by their angles. For example, a **right triangle** is a triangle with a right angle (square corner) of 90°.

right triangle

Note: In a right triangle, the side opposite the right angle is called the **hypotenuse** and is always the longest side of the right triangle.

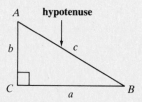

An **acute triangle** is a triangle whose three angles are all acute angles (less than 90°).

acute triangle

An **obtuse triangle** is a triangle with an angle that is an obtuse angle (more than 90°).

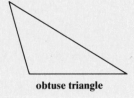

obtuse triangle

CHAPTER 2 PRACTICE

For each exercise 1–8, use a ruler to measure the length of the following line segment *AB* to the nearest:

A ——————————————— B

[See Example 1.]

1. inch   **2.** half-inch

3. quarter-inch   **4.** eighth-inch

5. sixteenth-inch   **6.** centimeter

7. millimeter   **8.** tenth-centimeter

For each exercise 9–14, convert the given U.S. measure to the indicated metric measure. [See Example 2.]

9. 2 in. = ? cm   **10.** $\frac{1}{2}$ m = ? mm   **11.** 6 ft = ? m

12. 1 ft = ? cm   **13.** 100 yd = ? m   **14.** 50 mi = ? km

For each exercise 15–20, convert the given metric measure to the indicated U.S. measure. [See the "Check" following Example 2.]

15. 1500 mm = ? in.   **16.** 85 cm = ? in.   **17.** 5 m = ? ft

18. 100 m = ? yd   **19.** 10 km = ? mi   **20.** 0.25 km = ? yd

For exercises 21–32, **a.** use a protractor to measure each angle **b.** identify each angle as an acute, an obtuse, a right, or a straight angle.

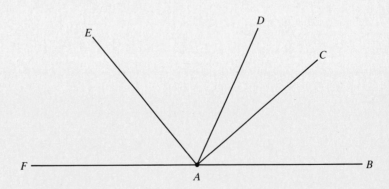

[See Example 3 and pages 24–25.]

21. $\angle BAC$	**22.** $\angle BAD$	**23.** $\angle BAE$
24. $\angle BAF$	**25.** $\angle CAD$	**26.** $\angle CAE$
27. $\angle CAF$	**28.** $\angle DAE$	**29.** $\angle DAF$
30. $\angle EAF$	**31.** $\angle FAB$	**32.** $\angle EDC$

For each exercise 33–44, rename the given angle measure in decimal form, as an equal amount of degrees. Round to the nearest hundredth degree when necessary.

33. $36°15'$	**34.** $25°30'$	**35.** $82°45'$
36. $7°57'$	**37.** $105°20'$	**38.** $93°40'$
39. $42°36''$	**40.** $75°18''$	**41.** $6°18'9''$
42. $102°21'27''$	**43.** $53°44'15''$	**44.** $19°52'30''$

For each exercise 45–50, rename the given degree measure in terms of degrees and minutes. [See the Note following Example 4.]

45. $17.5°$	**46.** $52.4°$	**47.** $46.75°$
48. $8.25°$	**49.** $103.35°$	**50.** $250.45°$

For each exercise 51–56, rename the given degree measure in terms of degrees, minutes, and seconds. [See the Note following Example 5.]

51. $39.11°$	**52.** $75.23°$	**53.** $130.51°$
54. $6.18°$	**55.** $82.59°$	**56.** $304.47°$

Copyright © 1990 by Harcourt Brace Jovanovich, Inc. All rights reserved.

For exercises 57–68, match each figure on the left with one and only one name on the right. Do this in such a way that no two figures have the same name. [See pages 18–19, and 24–28.]

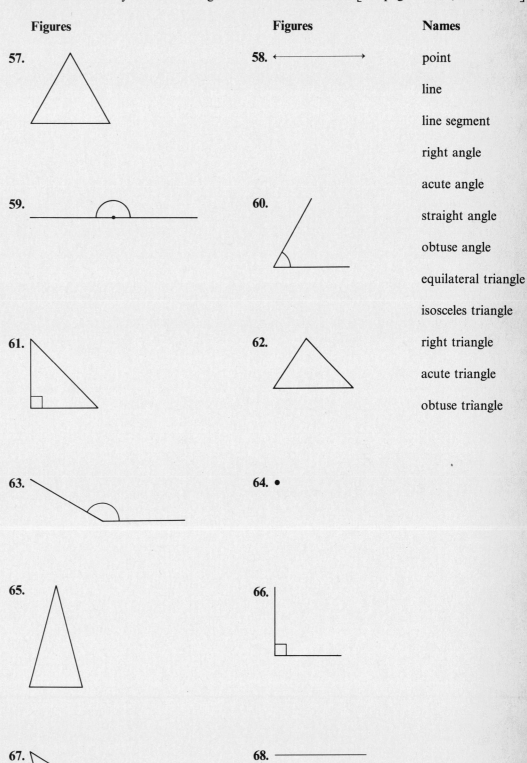

Figures	**Figures**	**Names**
57.	**58.**	point
		line
		line segment
		right angle
		acute angle
59.	**60.**	straight angle
		obtuse angle
		equilateral triangle
		isosceles triangle
61.	**62.**	right triangle
		acute triangle
		obtuse triangle
63.	**64.**	
65.	**66.**	
67.	**68.**	

Copyright © 1990 by Harcourt Brace Jovanovich, Inc. All rights reserved.

CHAPTER 3

PARALLEL AND PERPENDICULAR LINES

In this chapter you will

- Apply properties of parallel lines.
- Apply properties of perpendicular lines.

3.1 PROPERTIES OF PARALLEL LINES

Straight lines that never intersect are called **parallel lines.** In geometry, the definition for parallel lines is often stated as follows. Two or more lines that lie in the same plane are **parallel lines** if they have no points in common. That is, parallel lines never intersect no matter how they are extended in the plane.

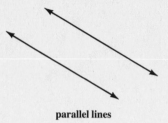

parallel lines

Note: Parallel lines are always the same distance apart, like railroad tracks.

Straight lines that intersect at exactly one point are called **intersecting lines.** In geometry, the definition for intersecting lines is often stated as follows. If two distinct lines in the same plane are not parallel, then they are intersecting lines.

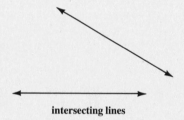

intersecting lines

In general, parallel lines are often denoted by letters. For example, the following lines l_1 and l_2 are parallel,

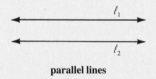

parallel lines

and are denoted by

$$l_1 \parallel l_2$$

which is read as "*l* one is parallel to *l* two."

When two or more parallel lines are intersected by a third line, the third line is called a **transversal.**

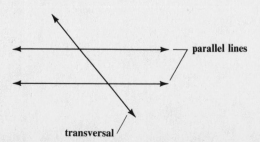

parallel lines

transversal

The eight angles formed when a transversal intersects two parallel lines are very important in geometry. For example:

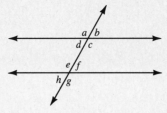

Figure 1

In Figure 1, the **supplementary angles** (angles whose sum is 180°) are:

$\angle a$	and	$\angle b$
$\angle b$	and	$\angle c$
$\angle c$	and	$\angle d$
$\angle d$	and	$\angle a$
$\angle e$	and	$\angle f$
$\angle f$	and	$\angle g$
$\angle g$	and	$\angle h$
$\angle h$	and	$\angle e$

supplementary angles

Note: Supplementary angles have a common vertex and a common side and always combine to form a straight line.

In Figure 1, the **vertical angles** (two angles opposite each other whose sides are the same two lines) are:

$\angle a$	and	$\angle c$
$\angle b$	and	$\angle d$
$\angle e$	and	$\angle g$
$\angle f$	and	$\angle h$

vertical angles

Note: Vertical angles are always congruent. For example, in Figure 1:

$$\angle a \cong \angle c$$
$$\angle b \cong \angle d$$
$$\angle e \cong \angle g$$
$$\angle f \cong \angle h$$

In Figure 1, the **alternate interior angles** (two angles that lie between the parallel lines and on opposite sides of the transversal) are:

$\angle c$	and	$\angle e$
$\angle d$	and	$\angle f$

alternate interior angles

Note: Alternate interior angles are always congruent. For example, in Figure 1:

$$\angle c \cong \angle e$$
$$\angle d \cong \angle f$$

In Figure 1, the **alternate exterior angles** (two angles that lie outside the parallel lines and on opposite sides of the transversal) are:

| $\angle a$ | and | $\angle g$ |
| $\angle b$ | and | $\angle h$ |

alternate exterior angles

Note: Alternate exterior angles are always congruent. For example in Figure 1:

$$\angle a \cong \angle g$$
$$\angle b \cong \angle h$$

In Figure 1, the **corresponding angles** (two angles that are in the same corresponding position with respect to the two parallel lines and the transversal) are:

$\angle a$	and	$\angle e$
$\angle b$	and	$\angle f$
$\angle c$	and	$\angle g$
$\angle d$	and	$\angle h$

corresponding angles

Note: Corresponding angles are always congruent. For example, in Figure 1:

$$\angle a \cong \angle e$$
$$\angle b \cong \angle f$$
$$\angle c \cong \angle g$$
$$\angle d \cong \angle h$$

> Given parallel lines and a transversal through them, you can find the measure of each angle formed (without using a protractor) by knowing the measure of just one of the angles formed.

Example 1 Find the measure of each angle in the following figure (without using a protractor), given that $l_1 \parallel l_2$ and $\angle a = 120°$.

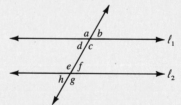

Solution Start by writing down the given angle and then make use of your knowledge of supplementary, vertical, alternate interior, and corresponding angles:

$\angle a = 120°$ is given.

$\angle b = 60°$ because $\angle a + \angle b = 180°$ (supplementary angles).

$\angle c = 120°$ because $\angle c \cong \angle a$ (vertical angles).

$\angle d = 60°$ because $\angle d \cong \angle b$ (vertical angles).

$\angle e = 120°$ because $\angle e \cong \angle a$ (corresponding angles) or
$\angle e \cong \angle c$ (alternate interior angles).

$\angle f = 60°$ because $\angle f \cong \angle b$ (corresponding angles) or
$\angle f \cong \angle d$ (alternate interior angles).

$\angle g = 120°$ because $\angle g \cong \angle e$ (vertical angles) or
$\angle g \cong \angle c$ (corresponding angles).

$\angle h = 60°$ because $\angle h \cong \angle f$ (vertical angles) or
$\angle h \cong \angle d$ (corresponding angles).   ■

Note: The angles formed when a transversal intersects parallel lines have at most two different measures, and the sum of those two measures is always 180°. For example, the angles formed in Example 1 were either 60° or 120°, and

$$60° + 120° = 180°$$

The following examples show how to solve the most common type of geometry problems involving one or more transversals intersecting parallel lines.

Example 2   Find the measure of $\angle x$ in the following figure (without using a protractor) given $l_1 \parallel l_2$.

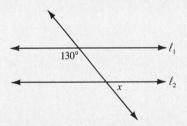

Solution   There are many different ways to solve this type of problem. For example:

- By the previous Note, $\angle x$ must either be 130° or its supplement, 50°. And

$$\angle x = 50°$$

because $\angle x$ is clearly an acute angle (less than 90°) in the given figure.

- Given the fact that vertical angles and alternate interior angles are congruent, $\angle x$ is a supplementary angle to an angle that measures 130°:

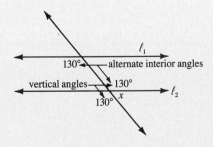

And so,

$$\angle x = 180° - 130°$$
$$= 50°$$

• Using supplementary angles, the angles adjacent to 130° are each 50°:

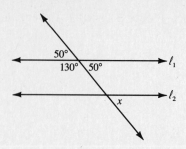

And so,

$$\angle x = 50°$$

because 50° and $\angle x$ are corresponding angles. ■

Example 3   Find the measure of $\angle n$ in the following figure (without using a protractor) given $l_1 \parallel l_2$.

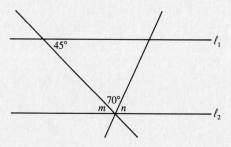

Solution   One way to solve this problem is to note that $\angle m$, 70°, and $\angle n$ form a straight angle

$$\angle m + 70° + \angle n = 180°$$

and $\angle m$ and 45° are congruent because they are alternate interior angles

$$\angle m = 45°$$

and so

$$45° + 70° + \angle n = 180°$$
$$115° + \angle n = 180°$$
$$\angle n = 180° - 115°$$
$$\angle n = 65° \quad ■$$

Parallel lines always have the same **slope.** Lines that rise from left to right have **positive slopes.** Lines that fall from left to right have negative slopes. Horizontal lines have a zero slope and vertical lines have undefined slopes.

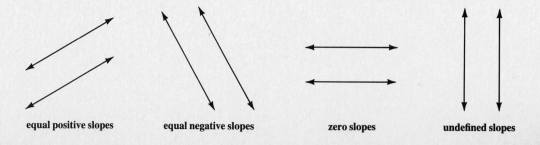

equal positive slopes   equal negative slopes   zero slopes   undefined slopes

3.2   USE PROPERTIES OF PERPENDICULAR LINES

Two lines are **perpendicular lines** if they intersect at right angles.

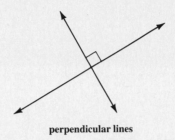

perpendicular lines

Perpendicular lines, if they are not vertical or horizontal, have slopes that are negative reciprocals of each other. And so, given two lines l_1 and l_2 that are not horizontal or vertical, if $l_1 \perp l_2$ (read as "l one is perpendicular to l two"), then

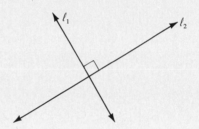

you have

$$\text{slope } l_1 = -\frac{1}{\text{slope } l_2}$$

and

$$\text{slope } l_2 = -\frac{1}{\text{slope } l_1}$$

or, put another way

$$(\text{slope } l_1)(\text{slope } l_2) = -1$$

Note: Horizontal lines have a slope of zero, and vertical lines have an undefined slope. And so the slopes of a horizontal and a vertical line are not negative reciprocals of each other, even though every horizontal line is perpendicular to every vertical line.

Example 4   If $l_1 \perp l_2$ and the slope of l_2 is $\frac{2}{3}$, what is the slope of l_1?

Solution   Because $l_1 \perp l_2$, you have

$$\text{slope } l_1 = -\frac{1}{\text{slope } l_2}$$

and so

$$\text{slope } l_1 = -\frac{1}{\frac{2}{3}} \qquad \text{Substitute.}$$

$$\text{slope } l_1 = -\frac{3}{2} \qquad \text{Simplify.} \quad \blacksquare$$

Note: In Example 4, the product of the slopes is -1, as it always will be for perpendicular lines that are not horizontal or vertical:

$$(\text{slope } l_1)(\text{slope } l_2) = \left(\frac{2}{3}\right)\left(-\frac{3}{2}\right) = -1$$

Example 5   If two lines have slope of $\frac{8}{3}$ and $-\frac{3}{8}$, are the two lines perpendicular?

Solution   Yes, the lines are perpendicular because the product of their slopes is -1:

$$\left(\frac{8}{3}\right)\left(-\frac{3}{8}\right) = -1 \quad \blacksquare$$

The following is an important statement about parallel and perpendicular lines.

When a transversal is perpendicular to one of two parallel lines that it intersects, it is also perpendicular to the other parallel line.

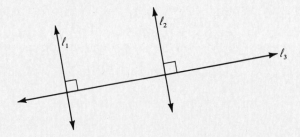

The previous statement is restated using symbols in the following box.

> If $l_1 \parallel l_2$ and $l_1 \perp l_3$, then $l_2 \perp l_3$.
>
> Or, equivalently:
>
> If $l_1 \perp l_3$ and $l_1 \parallel l_2$, then $l_2 \perp l_3$.

Example 6   If $l_1 \parallel l_2$ and $l_1 \perp l_3$ and the slope of l_3 is $-\frac{3}{4}$, what is the slope of l_2?

Solution   Start by drawing a diagram:

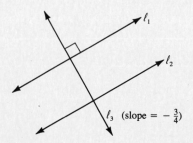

Because $l_1 \parallel l_2$ and $l_1 \perp l_3$, you have

$$l_2 \perp l_3$$

Because the product of the slopes of perpendicular lines is always -1, you have

$$(\text{slope } l_2)(\text{slope } l_3) = -1$$

and so

$$(\text{slope } l_2)\left(-\frac{3}{4}\right) = -1 \qquad \text{Substitute.}$$

$$\text{slope } l_2 = \frac{-1}{-\frac{3}{4}} \qquad \text{Solve for "slope } l_2\text{."}$$

$$\text{slope } l_2 = \frac{4}{3} \qquad \text{Simplify.} \quad \blacksquare$$

CHAPTER 3 PRACTICE

For each exercise 1–18, write "true" or "false."

1. Parallel lines never intersect.

2. Parallel lines always have the same slope.

3. Intersecting lines are never parallel.

4. Perpendicular lines are intersecting lines.

5. The notation $a \parallel b$ is read as "a is parallel to b."

6. The notation $a \perp b$ is read as "a is perpendicular to b."

7. A transversal is a line that intersects parallel lines.

8. When a transversal intersects two parallel lines, the angles formed are all congruent.

9. Supplementary angles have a sum of 180°.

10. Vertical angles are always congruent.

11. Alternate interior angles are always congruent.

12. Corresponding angles are always congruent.

13. Perpendicuar lines intersect at right angles.

14. Perpendicular lines always have slopes that are negative reciprocals of each other.

15. For perpendicular lines that are not horizontal or vertical, the product of their slopes is always −1.

16. When a transversal is perpendicular to one of two parallel lines that it intersects, it is also perpendicular to the other parallel line.

17. If $l_1 \parallel l_2$ and $l_1 \perp l_2$, then $l_2 \perp l_3$.

18. If $l_1 \perp l_3$ and $l_1 \parallel l_2$, then $l_2 \parallel l_3$.

For each exercise 19–30, use the following diagram (Figure 2) of two parallel lines intersected by a transversal.

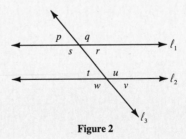

Figure 2

In Figure 2:

19. Which lines are parallel?

20. Which lines have the same slope?

Copyright © 1990 by Harcourt Brace Jovanovich, Inc. All rights reserved.

21. Which line is the transversal?

22. Which line has a different slope than the other two?

23. Which lines, if any, appear to have positive slopes?

24. Which lines, if any, appear to have negative slopes?

25. Which angles are vertical angles?

26. Which angles are supplementary angles?

27. Which angles are corresponding angles?

28. Which angles are alternate interior angles?

[See Example 1.]

29. Find the measure of each angle in Figure 2 if $\angle q = 135°$.

30. Find the measure of each angle in Figure 2 if $\angle t = 55°$.

[See Example 2.]

31. In the figure below, find the measure of $\angle a$ given that $l_1 \| l_2$.

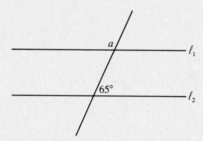

32. In the figure below, find $\angle m$ given that $l_1 \| l_2$.

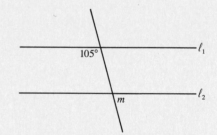

[See Example 3.]

33. In the figure below, find the measure of $\angle a$ given that $l_1 \| l_2$.

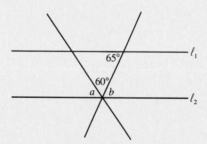

34. In the figure below, find the measure of $\angle y$ given that $l_1 \| l_2$.

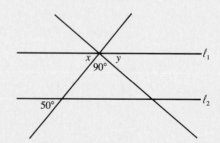

Copyright © 1990 by Harcourt Brace Jovanovich, Inc. All rights reserved.

[See Example 4.]

35. If $l_1 \perp l_2$ and the slope of l_2 is $\frac{3}{4}$, what is the slope of l_1?

36. If $l_1 \perp l_2$ and the slope of l_1 is $-\frac{5}{2}$, what is the slope of l_2?

[See Example 5.]

37. If two lines have slopes $\frac{2}{3}$ and $-\frac{3}{2}$, are the two lines perpendicular?

38. Are two lines with slopes 2 and $-\frac{1}{2}$ perpendicular?

[See Example 6.]

39. If $l_1 \parallel l_2$ and $l_1 \perp l_3$ and the slope of l_3 is $\frac{6}{5}$, what is the slope of l_2?

40. If $l_2 \perp l_3$ and $l_1 \parallel l_3$ and the slope of l_1 is $-\frac{1}{2}$, what is the slope of l_2?

Mixed Practice For each exercise 41–58, use the following Figure 3.

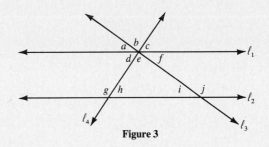

Figure 3

If in Figure 3 $l_1 \parallel l_2$ and $l_3 \perp l_4$, then:

41. Find $\angle a$ if $\angle f = 35°$.

42. Find $\angle d$ if $\angle c = 65°$.

43. Find $\angle d$ if $\angle h = 45°$.

44. Find $\angle i$ if $\angle f = 50°$.

45. Find $\angle c$ if $\angle h = 60°$.

46. Find $\angle i$ if $\angle a = 30°$.

47. Find $\angle g$ if $\angle h = 70°$.

48. Find $\angle i$ if $\angle j = 125°$.

49. Find $\angle a$ if $\angle c = 65°$.

50. Find $\angle h$ if $\angle i = 68°$.

51. Find $\angle a$ if $\angle j = 120°$.

52. Find $\angle g$ if $\angle c = 63°$.

53. Find $\angle c$ if $\angle i = 80°$.

54. Find $\angle f$ if $\angle h = 49°$.

55. Find the slope of l_1 if the slope of l_2 is $\frac{4}{5}$.

56. Find the slope of l_2 if the slope of l_1 is $-\frac{5}{3}$.

Copyright © 1990 by Harcourt Brace Jovanovich, Inc. All rights reserved.

57. Find the slope of l_3 if the slope of l_4 is $\frac{4}{5}$ and $l_3 \perp l_4$.

58. Find the slope of l_4 if $l_3 \perp l_4$ and the slope of l_3 is $-\frac{5}{3}$.

Extra For each exercise 59 and 60, draw a diagram before trying to answer the question.

59. If l_1, l_2, and l_3 are all parallel and $l_2 \perp l_4$ and the slope of l_3 is $\frac{5}{2}$, what is the slope of l_4?

60. If $l_1 \perp l_2$, $l_2 \perp l_3$, and $l_3 \perp l_4$ and the slope of l_1 is $\frac{3}{7}$, what is the slope of l_4?

Copyright © 1990 by Harcourt Brace Jovanovich, Inc. All rights reserved.

CHAPTER 4

PROPERTIES OF TRIANGLES

In this chapter you will

- Use properties of interior angles of a triangle.
- Use the Pythagorean Theorem.

| 4.1 | USE PROPERTIES OF INTERIOR ANGLES |

Recall: Every triangle has three sides, three vertices, and three angles.

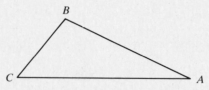

If you use a protractor to measure the three **interior angles** in $\triangle ABC$ above, you will find:

interior angles

$$\angle A = 25°$$
$$\angle B = 105°$$
$$\angle C = 50°$$

Verify the measures of $\angle A$, $\angle B$, and $\angle C$ using a protractor.

Note: The sum of the interior angles of $\triangle ABC$ is 180°:

$$\angle A + \angle B + \angle C = \mathbf{25° + 105° + 50°}$$
$$= 180°$$

The fact that the sum of the interior angles of $\triangle ABC$ is 180° is not a coincidence, as stated in the following box.

> The sum of the interior angles of any triangle is 180°.

When the measure of any two interior angles of a given triangle are known, you can always find the measure of the third interior angle by subtracting the sum of the two known measures from 180°.

Example 1   In $\triangle ABC$ below, that is the measure of angle $\angle C$?

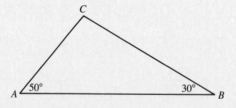

Solution   Start by finding the sum of the measures of the two known interior angles:

$$\angle A + \angle B = \mathbf{50° + 30°}$$
$$= 80° \longleftarrow \text{ sum}$$

Then subtract the sum from 180° to find the measure of angle C:

$$\angle C = 180° - 80°$$
$$= 100°$$

In $\triangle ABC$, the measure of angle C is 100°.

Check: In $\triangle ABC$, is the sum of the interior angles 180°? Yes:

$$\angle A + \angle B + \angle C = \mathbf{50° + 30° + 100°}$$
$$= 180° \longleftarrow 100° \text{ checks} \quad \blacksquare$$

Recall: A right triangle is a triangle with a right angle (90°).

Because the sum of the three interior angles of any triangle is 180° and a right triangle always has a right angle of 90°, the sum of the other two angles in a right triangle must always equal 90°:

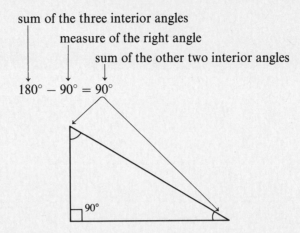

sum of the three interior angles

measure of the right angle

sum of the other two interior angles

$$180° - 90° = 90°$$

90°

Note: In a right triangle, the two interior angles that are not the right angle are both acute angles (less than 90°) because their sum is exactly 90°. And so:

> In a right triangle, you only need to know the measure of just one of the two acute angles in order to find the measure of the other acute angle.

Example 2   In the right triangle below, find $\angle B$.

A

60°

C   B

Solution   Start by finding the sum of the two known angles:

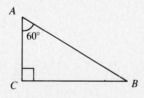

measure of a right angle

$$\angle A + \angle C = \mathbf{60° + 90°}$$
$$= 150° \longleftarrow \text{sum}$$

Then subtract that sum from 180°:

$$\angle B = 180° - 150°$$
$$= 30° \quad \blacksquare$$

Recall: An isosceles triangle is a triangle with two congruent sides and the base angles opposite the congruent sides are also congruent.

isosceles triangle

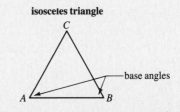

C

base angles

A   B

$$AC \cong BC \text{ (congruent sides)}$$
$$\angle A \cong \angle B \text{ (congruent base angles)}$$

In an isosceles triangle, you need only know the measure of just one angle in order to find the measures of the other two angles.

When given the measure of one of the two equal base angles in an isosceles triangle, you can find the measure of the other two angles as shown in Example 3.

Example 3   In the isosceles triangle below, find $\angle Y$ and $\angle Z$.

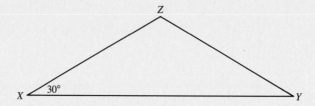

Solution   In an isosceles triangle, the base angles are congruent. And so,

$$\angle Y = \angle X = 30°$$

The sum of the base angles is 60°

$$\angle X + \angle Y = \mathbf{30° + 30°}$$
$$= 60°$$

and so,

$$\angle Z = 180° - 60°$$
$$= 120° \quad \blacksquare$$

When given the measure of the angle that is not one of the two equal base angles in an isosceles triangle, you can find the measure of the other two angles as shown in Example 4.

Example 4   In the isosceles triangle below, find $\angle B$ and $\angle C$.

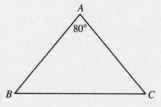

Solution   Start by finding the sum of the base angles:

$$\angle B + \angle C = 180° - \angle A$$
$$= 180° - \mathbf{80°} \qquad \text{Substitute.}$$
$$= 100° \longleftarrow \text{sum of base angles}$$

Then divide the sum of the two equal base angles by 2 to find the measure of the base angle:

$$\angle B = \angle C = 100° \div 2 = 50° \quad \blacksquare$$

Recall: An equilateral triangle is a triangle with all three sides congruent. Also recall that all three angles are congruent in an equilateral triangle.

In any equilateral triangle, each interior angle is 60° because each angle is congruent and

$$60° + 60° + 60° = 180°$$

Example 5 In the equilateral triangle below, find $\angle D$, $\angle E$, and $\angle F$.

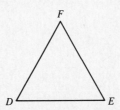

Solution In any equilateral triangle, the three congruent angles always measure 60° each. And so,

$$\angle D = \angle E = \angle F = 60° \quad ∎$$

4.2 USE THE PYTHAGOREAN THEOREM

Recall: In a right triangle, the side opposite the right angle is the longest side and is called the **hypotenuse.** The other two sides, which form the right angle, are called **legs.** The letters a and b are usually used to denote the legs of a right triangle and the letter c is used to denote the hypotenuse.

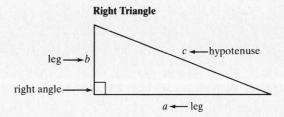

The following Pythagorean Theorem is one of the most important theorems in all of mathematics.

PYTHAGOREAN THEOREM
In any right triangle

$$a^2 + b^2 = c^2$$

where a and b are the lengths of the legs, and c is the length of the hypotenuse.

CAUTION The Pythagorean Theorem is true *only* for right triangles.

Note: In any right triangle:

The sum of the squares of the two legs always equals the square of the hypotenuse.

For example, the triangle shown below is a right triangle

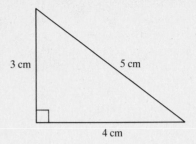

because

$$\underbrace{\text{legs} \qquad\qquad}_{} \qquad \underbrace{\text{hypotenuse}}_{}$$
$$3^2 + 4^2 = 9 + 16 = 25 = 5^2$$

Verify that the previous triangle is a right triangle with sides of 3 cm, 4 cm, 5 cm by measuring the length of each side with a ruler and measuring the right angle with a protractor.

The previous example of the Pythagorean Theorem

$$3^2 + 4^2 = 5^2$$

is interpreted geometrically in the following figure:

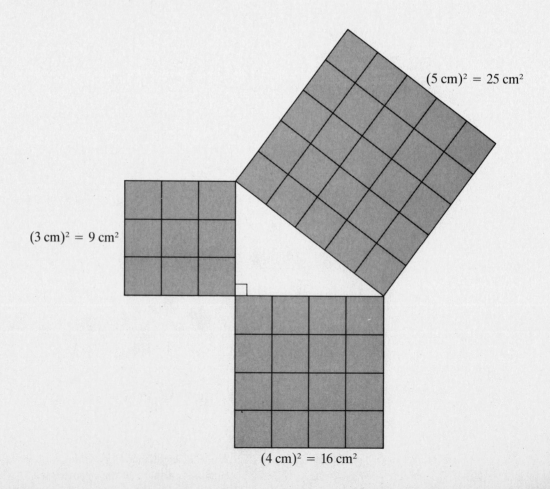

Verify the previous geometric interpretation by counting and adding the number of square centimeters on both legs of the right triangle and then counting the number of square centimeters on the hypotenuse.

To find the length of one side of a right triangle, given the lengths of the other two sides, you first substitute into the Pythagorean Theorem and then solve for the indicated letter, a, b, or c. For example, to find the length of the hypotenuse in the following right triangle

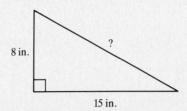

you use the Pythagorean Theorem as follows:

$$a^2 + b^2 = c^2$$
$$\mathbf{8}^2 + \mathbf{15}^2 = c^2 \qquad \text{Substitute.}$$
$$64 + 225 = c^2 \qquad \text{Solve for } c.$$
$$289 = c^2$$
$$c = \sqrt{289}$$
$$c = 17 \text{ (in.)}$$

To find the length of the leg in the following right triangle

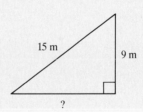

you use the Pythagorean Theorem as follows:

$$a^2 + b^2 = c^2$$
$$a^2 + \mathbf{9}^2 = \mathbf{15}^2 \qquad \text{Substitute.}$$
$$a^2 + 81 = 225 \qquad \text{Solve for } a.$$
$$a^2 = 144$$
$$a = \sqrt{144}$$
$$a = 12 \text{ (m)}$$

Recall: One way to tell if two given lines are perpendicular is to multiply slopes (assuming they are given and defined) to see if you get -1. For example, a line with slope $\frac{8}{3}$ is perpendicular to a line with slope $-\frac{3}{8}$ because

$$\left(\frac{8}{3}\right)\left(-\frac{3}{8}\right) = -1$$

Another way to tell if two lines are perpendicular is to use the converse of the Pythagorean Theorem.

CONVERSE OF THE PYTHAGOREAN THEOREM

If a, b, and c are the lengths of the three sides of a triangle and

$$a^2 + b^2 = c^2$$

then the triangle is a right triangle

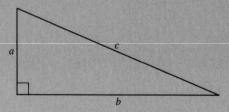

where a and b are the length of the legs, and c is the length of the hypotenuse.

Note: One consequence of the converse of the Pythagorean Theorem is that, if

$$a^2 + b^2 = c^2$$

then

$$a \perp b$$

Example 6   If a triangle has sides of 5 m, 12 m, and 13 m, are two of the sides perpendicular?

Solution   A triangle with sides of 5 m, 12 m, and 13 m is a right triangle because 5, 12, and 13 satisfy the Pythagorean Theorem:

$$
\begin{aligned}
a^2 + b^2 &= c^2 &&\longleftarrow \text{Pythagorean Theorem} \\
\mathbf{5}^2 + \mathbf{12}^2 &= \mathbf{13}^2 &&\text{Substitute.} \\
25 + 144 &= 169 &&\text{Simplify.} \\
169 &= 169 &&\longleftarrow \text{identity}
\end{aligned}
$$

And so in the given triangle, the sides with lengths 5 m and 12 m are perpendicular to each other:

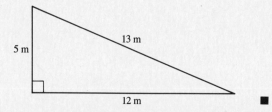

Note: There are no two sides that are perpendicular in a triangle with sides 2 m, 3 m, and 4 m because 2, 3, and 4 do not satisfy the Pythagorean Theorem:

$$
\begin{aligned}
a^2 + b^2 &= c^2 &&\longleftarrow \text{Pythagorean Theorem} \\
\mathbf{2}^2 + \mathbf{3}^2 &= \mathbf{4}^2 &&\text{Substitute.} \\
4 + 9 &= 16 &&\text{Simplify.} \\
13 &= 16 &&\longleftarrow \text{contradiction}
\end{aligned}
$$

And so in the given triangle, the sides with lengths 2 m and 3 m are not perpendicular to each other:

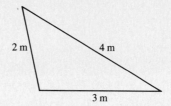

To find a right triangle in which the sides 2 m and 3 m are perpendicular to each other, you can use the Pythagorean Theorem.

Example 7   Find a right triangle with legs of 2 m and 3 m.

Solution   Start by drawing a right triangle and labeling its sides:

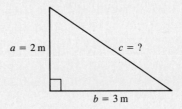

Then use the Pythagorean Theorem to find the required length of the hypotenuse:

$$a^2 + b^2 = c^2 \quad \longleftarrow \quad \text{Pythagorean Theorem}$$
$$2^2 + 3^2 = c^2 \qquad \text{Substitute.}$$
$$4 + 9 = c^2 \qquad \text{Solve for } c.$$
$$13 = c^2$$
$$c = \sqrt{13}$$

And so, the right triangle with legs of 2 m and 3 m must have a hypotenuse with length $\sqrt{13}$ m, or approximately 3.6 m.

Check: Does 2, 3, and $\sqrt{13}$ satisfy the Pythagorean Theorem? Yes:

$$a^2 + b^2 = c^2$$
$$2^2 + 3^2 = (\sqrt{13})^2 \qquad \text{Substitute.}$$
$$4 + 9 = 13 \qquad \text{Simplify.}$$
$$13 = 13 \quad \longleftarrow \quad \text{identity} \qquad \blacksquare$$

CHAPTER 4 PRACTICE

For each exercise 1–12, write "true" or "false."

1. The sum of the interior angles of any triangle is 180°.

2. When the measure of any two interior angles in a triangle are known, you can always find the measure of the third interior angle.

3. In any right triangle, the sum of the two acute angles is 90°.

4. In any right triangle, you must know the measure of both acute angles in order to find the measure of the third angle.

5. When the measure of just one angle in an isosceles triangle is known, you can always find the measures of the other two angles.

6. In any equilateral triangle, each interior angle is 60°.

7. In a right triangle with legs a and b and hypotenuse c,

$$a^2 + b^2 = c^2$$

8. If the sides of a triangle a, b, and c satisfy

$$a^2 + b^2 = c^2$$

then the triangle is a right triangle.

9. A triangle with sides of 1 m, 1 m, and 2m is a right triangle.

10. A triangle with sides of 1 m, 1 m, and $\sqrt{2}$ m is a right triangle.

11. If a and b are the legs of a right triangle then $a \perp b$.

12. If the sides of a triangle satisfy

$$a^2 + b^2 = c^2$$

then $a \perp b$.

For exercises 13–18, find the degree measure of each angle that is not given.

[See Example 1.]

[See Example 2.]

13.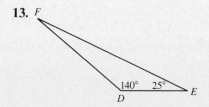

14.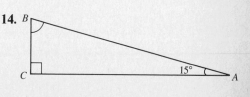

Copyright © 1990 by Harcourt Brace Jovanovich, Inc. All rights reserved.

[See Example 3.]

15. isosceles triangle

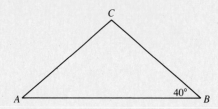

[See Example 4.]

16. isosceles triangle

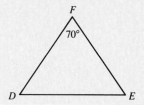

[See Example 5.]

17. equilateral triangle

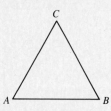

[See Examples 2 and 4.]

18. isosceles-right triangle

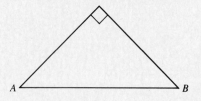

19. For the right triangle below, find the length of the hypotenuse.

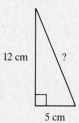

20. For the right triangle below, find the length of the indicated leg.

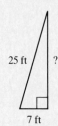

[See Example 6.]

21. If a triangle has sides of 6 cm, 8 cm, and 10 cm, are two of the sides perpendicular?

22. If a triangle has sides of 4 ft, 5 ft, and 6 ft, are two of the sides perpendicular?

[See Example 7.]

23. Find the right triangle with legs of 4 ft and 5 ft.

24. Find a right triangle with legs of 1 m and 2 m.

Copyright © 1990 by Harcourt Brace Jovanovich, Inc. All rights reserved.

Mixed Practice For exercises 25–36, complete the following table by finding the degree measure of each angle that is not given.

	Type of Triangle	∠A	∠B	∠C
25.	scalene	25°	92°	
26.	scalene		74°	105°
27.	isosceles	base angle 50°		
28.	isosceles			not a base angle 72°
29.	equilateral			
30.	right-isosceles			
31.	right	30°		
32.	right		45°	
33.	acute	37°		86°
34.	acute	89°	75°	
35.	obtuse		103°	25°
36.	obtuse	126°		53°

For exercises 37–48, complete the following table by finding the length of each leg or hypotenuse that will form a right triangle.

	leg *a*	leg *b*	hypotenuse *c*
37.	9 in.	12 in.	
38.	20 ft	21 ft	
39.	24 yd		26 yd
40.	12 mm		37 m

Copyright © 1990 by Harcourt Brace Jovanovich, Inc. All rights reserved.

	leg *a*	leg *b*	hypotenuse *c*
41.		40 cm	41 cm
42.		84 m	85 m
43.	1 in.	$\sqrt{2}$ in.	
44.	1 in.		$\sqrt{2}$ in.
45.	$\sqrt{10}$ ft	$\sqrt{15}$ ft	
46.	$6\sqrt{2}$ mm		$10\sqrt{2}$ m
47.	3.3 cm	5.6 cm	
48.	$4\frac{1}{2}$ yd	$2\frac{4}{5}$ yd	

Applications For each problem 49–54, first draw a right triangle, then label each known dimension (leg or hypotenuse), and then find the unknown dimension using the Pythagorean Theorem.

49. A guy wire reaches from the top of a vertical 60-foot flag pole to a ground anchor. The base of the flag pole is 25 feet from the anchor. How long is the guy wire?

50. A 17-foot ladder is leaning against the wall of a building. The base of the building is 8 feet from the base of the ladder. How high on the building does the ladder reach?

51. A rectangular field is 80 m long and 60 m wide. What is the minimum distance from one corner to the corner diagonally opposite?

52. Two vertical poles are 38 m and 46 m high, respectively. The bases of the poles are 15 m apart. How far is it from the top of one pole to the top of the other?

53. A pilot wants to fly 400 km due west. He takes the wrong course and flies in a straight line to end up 90 km due south of his planned destination. How far did the pilot fly?

54. A certain building is 3 miles west and 4 miles north of another building. Find the helicopter flying distance between the two buildings.

55. A triangle has sides of 2, 3, and *x* feet. For what two values of *x* will the triangle be a right triangle? [*Hint:* The length "*x* feet" could be a leg or the hypotenuse of the right triangle.]

56. A triangle has sides of 3, 8, and *x* inches. For what two values of *x* will the triangle be a right triangle?

Copyright © 1990 by Harcourt Brace Jovanovich, Inc. All rights reserved.

CONGRUENT AND SIMILAR TRIANGLES

In this chapter you will

- Identify congruent and similar triangles.
- Apply properties of similar triangles.

5.1   IDENTIFY CONGRUENT AND SIMILAR TRIANGLES

The concepts of *congruent* and *similar triangles* play a major role in the study of geometry. In **congruent triangles,** all three pairs of corresponding angles and sides are congruent; whereas, in **similar triangles,** only the three pairs of corresponding angles need be congruent.

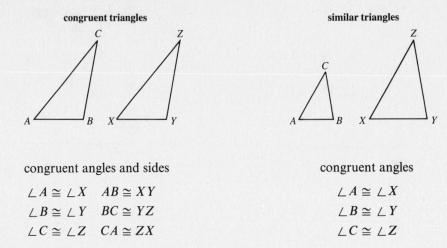

congruent angles and sides	congruent angles
$\angle A \cong \angle X$ $AB \cong XY$	$\angle A \cong \angle X$
$\angle B \cong \angle Y$ $BC \cong YZ$	$\angle B \cong \angle Y$
$\angle C \cong \angle Z$ $CA \cong ZX$	$\angle C \cong \angle Z$

Note: To indicate that two triangles *ABC* and *XYZ* are congruent, you write

$$\triangle ABC \cong \triangle XYZ$$

which is read as "triangle *ABC* is congruent to triangle *XYZ*."

Note: Congruent triangles are always similar triangles, but similar triangles are not necessarily congruent triangles. That is, congruent triangles always have exactly the same shape and size, whereas, similar triangles always have the same shape, but not necessarily the same size.

Because congruent triangles have exactly the same shape and size, they can be placed one on top of the other in such a way that they *fit perfectly* (corresponding sides and angles match exactly).

It is not necessary to know that all three pairs of corresponding angles and sides are congruent in order to guarantee that two triangles are congruent. For example, two triangles are congruent if they satisfy any one of the following four basic conditions for congruence.

Two triangles are congruent:
1. If all three pairs of corresponding sides are congruent, denoted by **SSS** (Side, Side, Side).

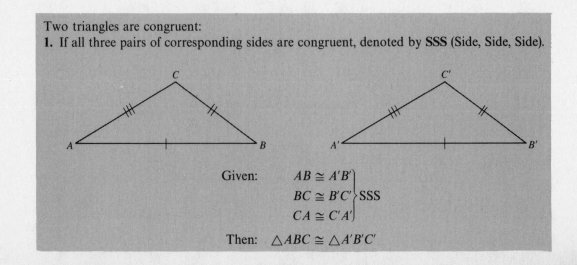

Given: $AB \cong A'B'$
            $BC \cong B'C'$ $\Big\}$ SSS
            $CA \cong C'A'$

Then: $\triangle ABC \cong \triangle A'B'C'$

2. If two pairs of corresponding sides and the angles between them are congruent, denoted by **SAS** (Side, Angle, Side).

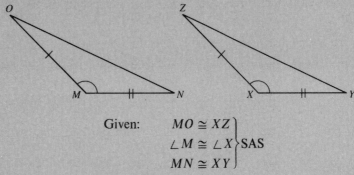

Given: $\left.\begin{array}{l} MO \cong XZ \\ \angle M \cong \angle X \\ MN \cong XY \end{array}\right\}$SAS

Then: $\triangle MNO \cong \triangle XYZ$

3. If two pairs of corresponding angles and the sides between them are congruent, denoted by **ASA** (Angle, Side, Angle).

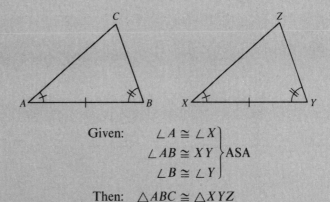

Given: $\left.\begin{array}{l} \angle A \cong \angle X \\ \angle AB \cong XY \\ \angle B \cong \angle Y \end{array}\right\}$ASA

Then: $\triangle ABC \cong \triangle XYZ$

4. If two pairs of corresponding angles and one pair of corresponding sides opposite one of the angles are congruent, denoted by **AAS** (Angle, Angle, Side).

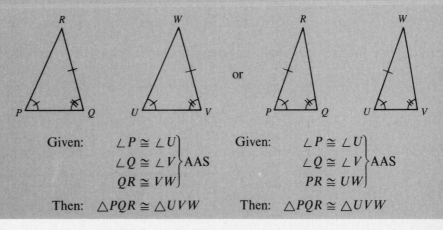

Given: $\left.\begin{array}{l} \angle P \cong \angle U \\ \angle Q \cong \angle V \\ QR \cong VW \end{array}\right\}$AAS

Then: $\triangle PQR \cong \triangle UVW$

Given: $\left.\begin{array}{l} \angle P \cong \angle U \\ \angle Q \cong \angle V \\ PR \cong UW \end{array}\right\}$AAS

Then: $\triangle PQR \cong \triangle UVW$

Note: You should convince yourself that each of the previous four sets of minimum conditions are enough to guarantee that the two triangles are congruent by constructing some triangles using the given conditions.

To determine whether two described triangles are congruent, you first use the given description to help draw the triangles and then test the given description, with the aid of your drawing, to see if it satisfies one of the four basic conditions for congruence; SSS, SAS, ASA, or AAS.

Example 1   Are two right triangles congruent if the two pairs of corresponding legs are congruent?

Solution   Start by drawing two right triangles using the given description:

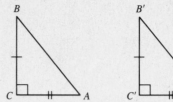

The two right triangles satisfy the third basic condition for congruence, SAS:

$$BC \cong B'C' \text{ (given)}$$
$$\angle C \cong \angle C' \text{ (right angles)}$$
$$AC \cong A'C' \text{ (given)}$$

And so, two right triangles are congruent if the two pairs of corresponding legs are congruent:

$$\triangle ABC \cong \triangle A'B'C' \quad \blacksquare$$

CAUTION   When drawing described triangles, be sure to consider all possible cases.

Example 2   Are two triangles congruent if all three pairs of corresponding angles are congruent?

Solution   Start by drawing triangles using the given description. There are two cases to consider:

Case 1 (corresponding sides congruent)   **Case 2** (corresponding sides not congruent)

In Case 1, the two triangles are congruent because they satisfy SSS:

$$AB \cong XY \text{ (by construction)}$$
$$BC \cong YZ \text{ (by construction)}$$
$$CA \cong ZX \text{ (by construction)}$$

In Case 2, the two triangles are not congruent because the corresponding pairs of sides are not congruent.

And so, two triangles are not necessarily congruent if all three pairs of corresponding angles are congruent.   ■

Note: The condition that all three pairs of corresponding angles are congruent, denoted by **AAA** (Angle, Angle, Angle), is not enough to guarantee that two triangles are congruent. However, the condition AAA is enough to guarantee that the two triangles are similar, by definition. That is, the sides, or "size," of the triangles has nothing to do with their being similar—similarity depends only on the angles, or "shape," of the triangles.

It is not necessary to know that all three pairs of corresponding angles are congruent in order to guarantee that two triangles are similar. Two triangles are similar if they satisfy any one of the following three basic conditions for similarity.

Two triangles are similar:

1. If only two pairs of corresponding angles are congruent. (See the following Note 1.)

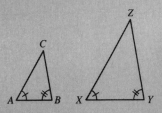

Given: $\angle A \cong \angle X$

$\angle B \cong \angle Y$

Then: $\triangle ABC$ is similar to $\triangle XYZ$.

2. If there are two vertical angles and the pair of corresponding sides opposite the vertical angles are parallel. (See the following Note 2.)

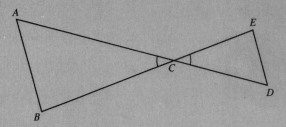

Given: $\angle DCE \cong \angle ACB$ (vertical angles)

$AB \parallel DE$

Then: $\triangle ABC$ is similar to $\triangle CDE$

3. If there are two corresponding angles (or common angles) formed by a transversal intersecting two parallel lines, and the corresponding sides opposite those angles are parallel. (See the following Note 3.)

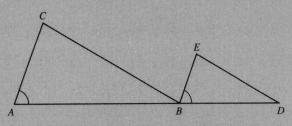
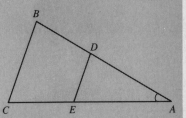

Given: $\angle A \cong \angle DBE$ (corresponding angles)

$AC \parallel BE$

Then: ABC is similar to $\triangle BDE$

Given: $\angle A \cong \angle A$ (common angle)

$BC \parallel DE$

Then: $\triangle ABC$ is similar to $\triangle ADE$

Note 1: The conditions

$$\angle A \cong \angle X$$

$$\angle B \cong \angle Y$$

are enough to guarantee that the two triangles ABC and XYZ are similar because the conditions are enough to guarantee that all three pairs of corresponding angles are congruent using: "The sum of the interior angles of a triangle always equals 180°."

Verify Note 1.

Note 2: The conditions

$$\angle DCE \cong \angle ACB$$
$$AB \parallel DE$$

are enough to guarantee that the two triangles ABC and CDE are similar because the conditions are enough to guarantee that all three corresponding angles are congruent using: "When a transversal intersects parallel lines, alternate interior angles are always congruent."

Verify Note 2.

Note 3: The conditions

$$\angle A \cong \angle DBE \qquad \text{or} \qquad \angle A \cong \angle A$$
$$AC \parallel BE \qquad\qquad\qquad BC \parallel DE$$

are enough to guarantee that the triangles ABC and BDE or triangles ABC and ADE, respectively, are similar because these conditions are enough to guarantee that all three pairs of corresponding angles are congruent using: "When a transversal intersects parallel lines, corresponding angles are always congruent."

Verify Note 3.

To determine whether two described triangles are similar, you first use the given description to help draw the triangles and then test the given description, with the aid of your drawing, to see if it satisfies one of the three basic conditions for similar triangles.

Example 3   Are two isosceles triangles similar if the corresponding non-base angles are formed by intersecting lines?

Solution   Start by drawing the two described isosceles triangles:

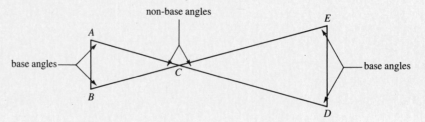

The two isosceles triangles ABC and CDE are similar because:

$$\angle BCA \cong \angle DCE \text{ (vertical angles)}$$
$$\angle A \cong \angle B \cong \angle C \cong \angle D \text{ (by definition of an isosceles}$$
$$\text{triangle and the fact that}$$
$$\angle A + \angle B = \angle D + \angle E) \quad \blacksquare$$

Note: If the non-base angles of two isosceles triangles are formed by intersecting lines, then the corresponding bases are parallel.

Verify the Note for AB and DE in the previous Example 3.

Example 4   Are two right triangles similar if they share a common acute angle?

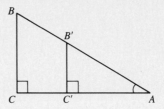

Solution   Start by drawing the right triangles using the description:

This problem can be solved using either condition 1 or condition 2 in the previous box:

By condition 1, the right triangles are similar because:

$$\angle A \cong \angle A \text{ (common angle)}$$
$$\angle C \cong \angle C' \text{ (right angles)}$$

By condition 2, the right triangles are similar because:

$$\angle A \cong \angle A \text{ (common angle)}$$
$$BC \parallel B'C' \text{ (both are perpendicular to } AC)$$

And so, two right triangles are similar if they share a common acute angle.   ■

Note: Two right triangles are not necessarily similar if the angle they share is the right angle.

Verify the Note.

CAUTION   When drawing described triangles, be sure to consider all possible cases.

Example 5   Are two isosceles triangles similar if one pair of corresponding base angles are congruent, or if one pair of corresponding non-base angles are congruent?

Solution   Start by drawing isosceles triangles using the given description. There are two cases to consider:

Case 1 (base angles congruent)   **Case 2** (non-base angles congruent)

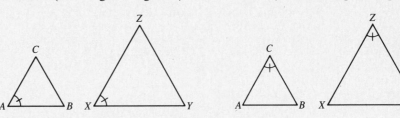

In Case 1, the isosceles triangles are similar because:

$$\angle A \cong \angle X \text{ (given)}$$
$$\angle B \cong \angle Y \text{ (by definition of an isosceles triangle)}$$
$$\angle C \cong \angle Z \text{ (by condition 1 in the previous box)}$$

In Case 2, the isosceles triangles are similar because:

$$\angle C \cong \angle Z \text{ (given)}$$
$$\angle A \cong \angle B \cong \angle X \cong \angle Y \text{ (by definition of an isosceles triangle and the fact that}$$
$$\angle A + \angle B = \angle X + \angle Y)$$

And so, two isosceles triangles are similar if one pair of corresponding base angles are congruent or if one pair of corresponding non-base angles are congruent. ■

Note: Two isosceles triangles are not necessarily similar if just two corresponding angles are congruent.

Verify the Note.

5.2 **USE PROPERTIES OF SIMILAR TRIANGLES**

Up to now you have studied how to identify congruent and similar triangles. In the remainder of this section, you will use these skills to help apply common properties of similar triangles. The first, and most important property of similar triangles is given in the following box.

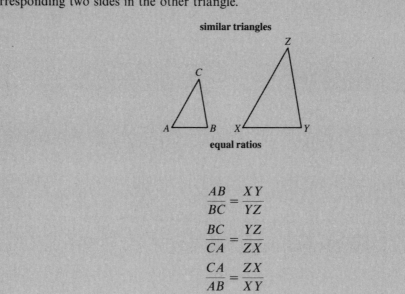

In two similar triangles, the ratio of any two sides in one triangle equals the ratio of the corresponding two sides in the other triangle.

similar triangles

equal ratios

$$\frac{AB}{BC} = \frac{XY}{YZ}$$

$$\frac{BC}{CA} = \frac{YZ}{ZX}$$

$$\frac{CA}{AB} = \frac{ZX}{XY}$$

Note: The three equal ratios in the previous box are all direct proportions.

To solve a problem using a proportion, such as

$$\frac{AB}{BC} = \frac{XY}{YZ}$$

you must know either three of the four terms (see the following Example 6), or know two of the four terms and be able to represent two others in terms of the same variable (see the following Example 7).

Example 6 Given similar triangles ABC and XYZ below, find the length of AB.

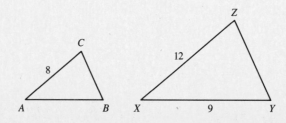

Solution   Start by using a variable, such as x, to represent the length of the desired side AB:

$$\text{Let } x = \text{the length of } AB.$$

Then use the fact that the given triangles are similar to write two equal ratios in such a way that x is the only unknown term:

$$\triangle ABC \qquad\qquad\qquad \triangle XYZ$$

$$\frac{x}{8} \quad\begin{matrix}\longleftarrow AB \text{ corresponds to } XY \longrightarrow\\ \longleftarrow AC \text{ corresponds to } XZ \longrightarrow\end{matrix}\quad \frac{9}{12}$$

And then use the two equal ratios to write and solve a direct proportion:

$$\frac{x}{8} = \frac{9}{12} \longleftarrow \text{ direct proportion}$$

$$12x = 8(9) \qquad \text{Solve for } x.$$

$$x = \frac{72}{12}$$

$$x = 6$$

And so, the length of AB is 6 units.   ■

Note: To find the length of AB in Example 6, you needed to write and solve a direct proportion using two equal ratios. This was possible to do because the given triangles were similar triangles.

CAUTION   If two triangles are not similar, then you will not, in general, be able to find the length of a given side using a direct proportion because the ratios of corresponding sides will not, in general, be equal.

When given a figure constructed of triangles and asked to find the length of the sides, the first thing to check is whether two of the triangles are similar. And, if they are similar, use equal ratios, if you can, to help find the length of the desired side.

Example 7   Given $BC \parallel DE$ in the figure below, find AE.

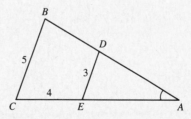

Solution   Start by identifying similar triangles:
The triangles ABC and ADE are similar because

$$\angle A \cong \angle A \text{ (common angle)}$$
$$BC \parallel DE \text{ (given)}$$

Then use a variable to represent the length of the desired side AE:

$$\text{Let } x = \text{the length of } AE.$$

Then use corresponding sides of the two similar triangles to write and solve a direct proportion:

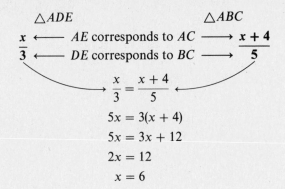

$$5x = 3(x + 4)$$
$$5x = 3x + 12$$
$$2x = 12$$
$$x = 6$$

And so, the length of AE is 6 units.　■

Note: Without the condition $BC \parallel DE$ in Example 7, you would not be able to find the length of AE. The given fact that $BC \parallel DE$ makes it possible to form two similar triangles. And, by doing so, to find the length of AE.

There are two types of special triangles that are used so often in building and engineering that it is worth taking the time and effort to learn and memorize their basic properties. The first of these two special triangles is a **45-45-90° triangle,** or equivalently, an **isosceles-right triangle:**

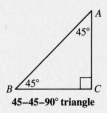

45–45–90° triangle

Note: If in a 45-45-90° triangle, the length of each leg is x units, then

$$\text{the length of the hypotenuse} = x\sqrt{2} \text{ units}$$

because x, x, and $x\sqrt{2}$ satisfies the Pythagorean Theorem:

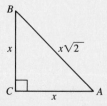

$$a^2 + b^2 = c^2 \longleftarrow \text{Pythagorean Theorem}$$
$$x^2 + x^2 = (x\sqrt{2})^2$$
$$2x^2 = 2x^2 \longleftarrow \text{identity}$$

The important properties of 45-45-90° triangles are summarized in the following box.

PROPERTIES OF 45-45-90° TRIANGLES

1. The angles opposite the legs are both 45°:

$$\angle A = \angle B = 45°$$

2. The legs are congruent:

$$AC = BC$$

3. The length of the hypotenuse is equal to the length of either leg times $\sqrt{2}$:

$$AB = AC\sqrt{2} \quad \text{and} \quad AB = BC\sqrt{2}$$

4. The length of each leg is equal to the length of the hypotenuse divided by $\sqrt{2}$:

$$AC = \frac{AB}{\sqrt{2}} \quad \text{and} \quad BC = \frac{AB}{\sqrt{2}}$$

Example 8   Given the 45-45-90° triangle below, find AC and AB.

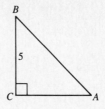

Solution   By property 2 of 45-45-90° triangles,

$$AC = BC = \mathbf{5} \text{ units}$$

By property 3 of 45-45-90° triangles,

$$AB = BC\sqrt{2} = \mathbf{5}\sqrt{2} \text{ units } (\approx 7.071 \text{ units}) \quad \blacksquare$$

Example 9   Given the isosceles-right triangle below, find $\angle D$ and EF.

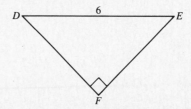

Solution   An isosceles-right triangle is a 45-45-90° triangle. And so, by property 1 of 45-45-90° triangles

$$\angle D = 45°$$

By property 4 of 45-45-90° triangles

$$EF = \frac{DE}{\sqrt{2}} = \frac{\mathbf{6}}{\sqrt{2}} = \frac{6}{\sqrt{2}} \cdot \frac{\sqrt{2}}{\sqrt{2}} = \frac{6\sqrt{2}}{2} \text{ units } (\approx 4.243 \text{ units}) \quad \blacksquare$$

Example 10   Given the square below, find the length of the diagonal.

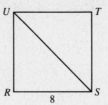

Solution   In the given square, both triangles RSU and STU are 45-45-90° triangles. And by property 4 of 45-45-90° triangles,

$$SU = \frac{RS}{\sqrt{2}} = \frac{8}{\sqrt{2}} = \frac{8\sqrt{2}}{2} = 4\sqrt{2} \text{ units}$$

And so, the length of the diagonal in the given square is $4\sqrt{2}$ units, or rounded to the nearest thousandth of a unit, 5.657 units.   ■

The second, and last special triangle to be considered in this section is a **30-60-90° triangle:**

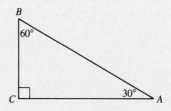

Note: If in a 30-60-90° triangle, the length of the leg opposite the 30°-angle is x units, then

$$\text{the length of the other leg} = x\sqrt{3} \text{ units}$$

and

$$\text{the length of the hypotenuse} = 2x \text{ units.}$$

Note: x, $x\sqrt{3}$, and $2x$ satisfy the Pythagorean Theorem:

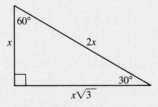

$$a^2 + b^2 = c^2 \longleftarrow \text{Pythagorean Theorem}$$
$$x^2 + (x\sqrt{3})^2 = (2x)^2$$
$$x^2 + 3x^2 = 4x^2 \longleftarrow \text{identity}$$

The important properties of 30-60-90° triangles are summarized in the following box.

PROPERTIES OF 30-60-90° TRIANGLES

1. The hypotenuse is twice as long as the leg opposite the 30°-angle. Or, put another way, the leg opposite the 30°-angle is one-half as long as the hypotenuse.

$$AB = 2BC$$

$$BC = \frac{AB}{2}$$

2. The leg opposite the 60°-angle equals the length of the other leg times $\sqrt{3}$. Or, put another way, the leg opposite the 30°-angle equals the length of the other leg divided by $\sqrt{3}$.

$$AC = BC\sqrt{3}$$

$$BC = \frac{AC}{\sqrt{3}}$$

Example 11 Given triangle XYZ below, find XY and XZ.

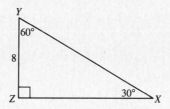

Solution By property 1 of 30-60-90° triangles,

$$XY = 2YZ = 2(8) = 16 \text{ units}$$

By property 2 of 30-60-90° triangles,

$$XZ = YZ\sqrt{3} = 8\sqrt{3} \text{ units } (\approx 13.856 \text{ units}) \quad \blacksquare$$

Example 12 Given triangle ABC below, find AC and BC.

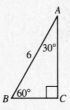

Solution By property 1 of 30-60-90° triangles,

$$BC = \frac{AB}{2} = \frac{6}{2} = 3 \text{ units}$$

By property 2 of 30-60-90° triangles,

$$AC = BC\sqrt{3} = 3\sqrt{3} \text{ units } (\approx 5.196 \text{ units}) \quad \blacksquare$$

Some geometry problems require both the properties of 45-45-90° triangles and 30-60-90° triangles.

Example 13   Given the figure below, find *AB* and *AD*.

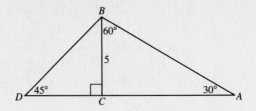

Solution   By property 1 of 30-60-90° triangles,

$$AB = 2BC = 2(5) = 10 \text{ units}$$

The length of *AD* equals the sum of the lengths of *AC* and *CD*:

$$AD = AC + CD$$

By property 2 of 30-60-90° triangles,

$$AC = BC\sqrt{3} = 5\sqrt{3} \text{ units}$$

By property 2 of 45-45-90° triangles,

$$CD = BC = 5 \text{ units}$$

And so,

$$AD = AC + CD = (5\sqrt{3} + 5) \text{ units } (\approx 13.660 \text{ units}) \quad \blacksquare$$

Note: The terms $5\sqrt{3}$ and 5 are not like terms and as such, cannot be combined.

CHAPTER 5 PRACTICE

For each exercise 1–12, write "true" or "false."

1. Two triangles are congruent if all three pairs of corresponding sides are congruent.

2. Two triangles are similar if all three pairs of corresponding angles are congruent.

3. Two triangles are similar if only two pairs of corresponding angles are congruent.

4. Two triangles are congruent if only two pairs of corresponding sides are congruent.

5. Congruent triangles are always similar triangles.

6. Similar triangles are always congruent triangles.

7. Similar triangles always have the same shape.

8. Congruent triangles always have the same size and shape.

9. Two right triangles are congruent if the two pairs of corresponding legs are congruent.

10. Two equilateral triangles are similar but not necessarily congruent.

11. Two 45-45-90° triangles are congruent.

12. Two 30-60-90° triangles are similar.

For each exercise 13–18, **a.** determine whether the given conditions are enough to guarantee that the two triangles are congruent and **b.** give the reason for your decision. See Examples 1 and 2.]

13. Given: $MN \cong XY$
$NO \cong YZ$
$OM \cong ZX$

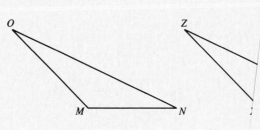

14. Given: $AB \cong A'B'$
$BC \cong B'C'$
$\angle B \cong \angle B'$

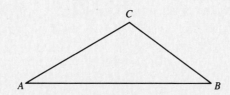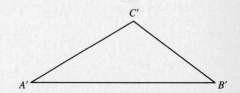

Copyright © 1990 by Harcourt Brace Jovanovich, Inc. All rights reserved.

15. Given: $\angle B \cong \angle Y$
          $\angle C \cong \angle Z$
          $BC \cong XY$

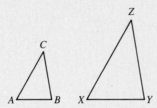

16. Given: $\angle P \cong \angle U$
          $\angle Q \cong \angle V$
          $PR \cong UW$

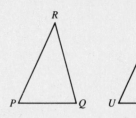

17. Given: $\angle A \cong \angle X$
          $\angle B \cong \angle Y$
          $\angle C \cong \angle Z$

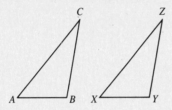

18. Given: right triangles ABC and $A'B'C'$
          $AC = A'C'$
          $BC = B'C'$

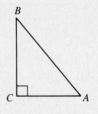
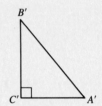

For each exercise 19–24 **a.** determine whether the given conditions are enough to guarantee that the two triangles are similar and **b.** give the reason for your decision. [See Examples 3–5.]

19. Given: $\angle A \cong \angle X$
          $\angle C \cong \angle Z$

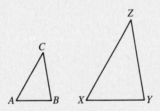

20. Given: triangles ABC and ADE
          $BC \parallel DE$

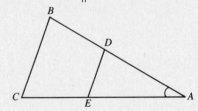

21. Given: right triangles ABC and ADE

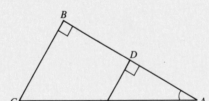

22. Given: right triangles ABC and $A'B'C$

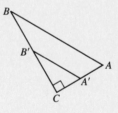

23. Given: triangles ABC and CDE
          $AB \parallel DE$

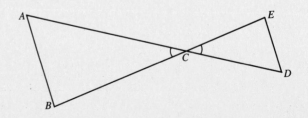

Copyright © 1990 by Harcourt Brace Jovanovich, Inc. All rights reserved.

24. Given: right triangles ABC and CDE
$AB \parallel CD$

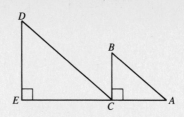

For exercises 25–28, find each indicated length using given information and the following similar triangles ABC and XYZ. [See Example 6.]

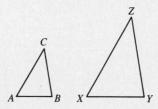

25. Given: $BC = 6$ m
$XY = 9$ m
$YZ = 5$ m
Find AB.

26. Given: $XY = 10$ cm
$AB = 8$ cm
$YZ = 15$ cm
Find BC.

27. Given: $AB = 15$ in.
$BC = 18$ in.
$CA = 20$ in.
$XY = 20$ in.
Find YZ and XZ.

28. Given: $XY = 8$ ft
$YZ = 12$ ft
$ZX = 15$ ft
$AC = 6$ ft
Find AB and BC.

For exercises 29–32, find each indicated length using the given information and $AB \parallel DE$ in the following triangles ABC and CDE. [See Example 6.]

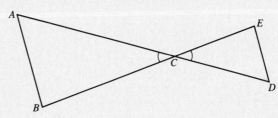

29. Given: $AB = 8$ km
$BC = 10$ km
$CE = 6$ km
Find DE.

30. Given: $AB = 10$ m
$EC = 15$ m
$DE = 8$ m
Find BC.

31. Given: $AB = 8$ mi
$BC = 12$ mi
$CA = 15$ mi
$CD = 10$ mi
Find DE, CE, AD, and BE.

32. Given: $CD = 30$ ft
$DE = 25$ ft
$EC = 35$ ft
$AB = 30$ ft
Find BC, AC, AD, and BE.

Copyright © 1990 by Harcourt Brace Jovanovich, Inc. All rights reserved.

For exercises 33–36, find each indicated length using the given information and the following right triangles ABC and $AB'C'$. [See Examples 6 and 7.]

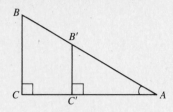

33. Given: $AB = 15$ m
                 $AC = 10$ m
                 $AB' = 8$ m
   Find AC'.

34. Given: $BC = 20$ in.
                 $AB' = 12$ in.
                 $AC' = 10$ in.
   Find AC.

35. Given: $BC = 50$ cm
                 $CC' = 15$ cm
                 $B'C' = 30$ cm
   Find AC'.

36. Given: $AC' = 5$ mi
                 $B'C' = 6$ mi
                 $BC = 10$ mi
   Find AC.

For exercises 37–40, find each indicated length using the given information and the following 45-45-90° triangle. [See Example 8.]

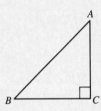

37. Given $BC = 3$ cm,
   find AC and AB.

38. Given $AC = 4$ yd,
   find AB and BC.

39. Given $AB = 7$ km,
   find AC and BC.

40. Given $BA = 6$ mi,
   find CB and CA.

For exercises 41–44, find each indicated length using the given information and the following isosceles-right triangle XYZ. [See Example 9.]

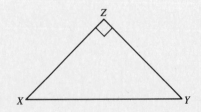

Copyright © 1990 by Harcourt Brace Jovanovich, Inc. All rights reserved.

41. Given $XZ = 5$ mi,
find YZ and XY.

42. Given $YZ = 8$ m,
find XY and XZ.

43. Given $XY = 2$ ft,
find XZ and YZ.

44. Given $YX = 12$ mm,
find ZY and ZX.

For exercises 45–48, find each indicated length using the given information and the following square $ABCD$. [See Example 10.]

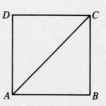

45. Given $AB = 2$ m,
find AC.

46. Given $AD = 5$ ft,
find AC.

47. Given $AC = 20$ mi,
find CD.

48. Given $AC = 10$ cm,
find BC.

For exercises 49–54, find each indicated length using the given information and the following 30-60-90° triangle. [See Examples 11 and 12.]

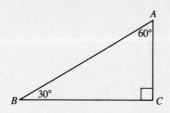

49. Given $AC = 4$ m,
find BC and AB.

50. Given $AC = 3$ yd,
find AB and BC.

51. Given $BC = 10$ ft,
find AC and AB.

52. Given $BC = 12$ cm,
find AB and AC.

53. Given $AB = 20$ km,
find AC and BC.

54. Given $AB = 9$ mi,
find BC and AC.

Copyright © 1990 by Harcourt Brace Jovanovich, Inc. All rights reserved.

For exercises 55–60, find each indicated length using the given information and the following figure. [See Example 13.]

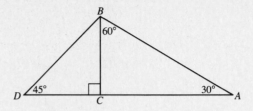

55. Given $CD = 4$ cm,
find BD, AC, and AB.

56. Given $AC = 6$ ft,
find AB, BD, and CD.

57. Given $AB = 9$ in.,
find AC, DC, and BD.

58. Given $BD = 12$ m,
find CD, AC, and AB.

59. Given $BC = 2$ km,
find AB, BD, and AD.

60. Given $BC = 3$ mi,
find AB, BD, and AD.

61. Given $AE \parallel BD$ in the figure below, find AB.

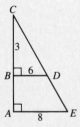

62. Given $BC \parallel B'C'$ in the figure below, find AC.

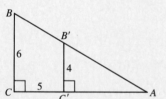

63. Given $XY \parallel BC$ in the figure below, find the ratio of XB to YC.

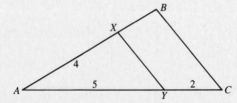

64. Given the triangle below, what is the exact measure of each angle?

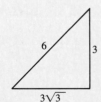

Copyright © 1990 by Harcourt Brace Jovanovich, Inc. All rights reserved.

Interesting Facts and Applications

65. A person who is 6 feet tall casts a shadow that is 5 feet long at the same time that a tree casts a 30-foot shadow. How tall is the tree?

66. If a surveyor took the measurements shown below, and if $AB \parallel DE$, then how long is the island, along line segment DE?

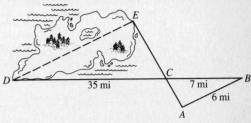

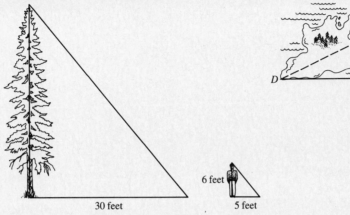

Use the following FACT for problems 67 and 68.

FACT The circular disk of a Susan B. Anthony dollar coin (1 inch diameter), when held at a distance of 108 inches (9 feet) from your eye, will appear to exactly block out the disk of the sun or moon.

67. What is the diameter of the sun to the nearest ten thousand miles given that the sun is about 93 million miles from the earth? (*Hint:* Use similar right triangles *ABC* and *BDE* and let

$$x = \text{one-half the diameter of the sun.}$$

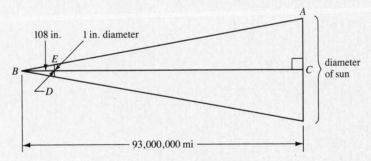

68. What is the distance from the earth to the moon to the nearest ten thousand miles given the diameter of the moon as about 2200 miles? (See the Hint for problem 67.)

Draw and label a figure below, similar to the figure in problem 67.

Copyright © 1990 by Harcourt Brace Jovanovich, Inc. All rights reserved.

CHAPTER 6

PERIMETER, AREA, AND VOLUME

In this chapter you will

- Identify common 2- and 3-dimensional geometric figures.
- Find the perimeter of a polygon.
- Find the circumference of a circle.
- Find the area of common polygons and circles.
- Find the volume of common 3-dimensional figures.

6.1   IDENTIFY COMMON GEOMETRIC FIGURES

Recall: A triangle is a polygon with three sides, three vertices, and three angles.

A **quadrilateral** is a polygon with four sides.

quadrilateral

Note: Every quadrilateral has four sides, four vertices, and four angles.

A **rectangle** is a quadrilateral with four right angles (square corners).

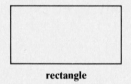

rectangle

Note: In a rectangle, opposite sides are always parallel.

A **square** is a rectangle with four sides that are all congruent.

square

Note: In a square, all four sides are congruent and all four angles are congruent (90° each).

A **parallelogram** is a quadrilateral with two pairs of parallel sides.

parallelogram

Note: In a parallelogram, opposite sides are always congruent and opposite angles are always congruent.

A **rhombus** is a parallelogram with four sides that are all congruent.

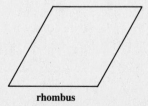

rhombus

Note: In a rhombus, the diagonals are always congruent and perpendicular to each other.

A **trapezoid** is a quadrilateral with one pair of parallel sides.

trapezoid

Note: In a trapezoid, it is possible that no two sides or angles are congruent. Also, every parallelogram is a trapezoid, but not every trapezoid is a parallelogram.

Note: In any polygon, the number of sides equals the number of angles. For example:

- A triangle has 3 sides and 3 angles.

- A quadrilateral has 4 sides and 4 angles.

- A **pentagon** has 5 sides and 5 angles.

- A **hexagon** has 6 sides and 6 angles.

- An **octagon** has 8 sides and 8 angles.

When the word "regular" is used to describe a polygon, as in *regular pentagon*, *regular hexagon*, or *regular octagon*, it means that all sides of the indicated polygon are congruent and also that all angles of the indicated polygon are congruent. For example, a **regular pentagon** is a pentagon with five sides that are congruent and five angles that are congruent.

regular pentagon

A **regular hexagon** is a hexagon with six sides that are congruent and six angles that are congruent. The cells in a beehive are regular hexagons.

regular hexagon

A **regular octagon** is an octagon with eight sides that are congruent and eight angles that are congruent. A STOP sign is a regular octagon.

regular octagon

A **regular decagon** is a decagon with ten sides that are congruent and ten angles that are congruent.

regular decagon

| 6.2 | **FIND THE PERIMETER** |

The total distance around a polygon is called the **perimeter** P. To find the perimeter of any polygon, you add the lengths of all the sides together.

Example 1   Find the perimeter of the following polygon using addition.

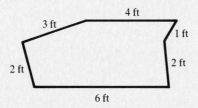

Solution   Add the lengths of all the sides together:

$$P = 2 \text{ ft} + 3 \text{ ft} + 4 \text{ ft} + 1 \text{ ft} + 2 \text{ ft} + 6 \text{ ft}$$
$$= 18 \text{ ft} \longleftarrow \text{ total distance around the polygon}$$

The perimeter of the polygon is 18 feet.   ■

When finding a perimeter, you will usually save time if you can use a formula. Most of the more common polygons have specific perimeter formulas.
The following are perimeter (P) formulas using length (l), width (w), and side (s):

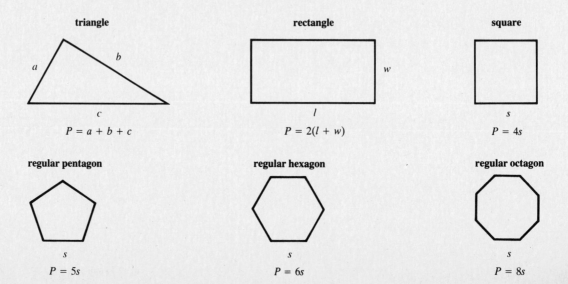

triangle	rectangle	square
$P = a + b + c$	$P = 2(l + w)$	$P = 4s$

regular pentagon	regular hexagon	regular octagon
$P = 5s$	$P = 6s$	$P = 8s$

To find the perimeter of a given figure using its perimeter formula, you write the correct perimeter formula, then substitute the known measures for the corresponding letters, and compute.

Example 2   Find the perimeter of the following square using a formula.

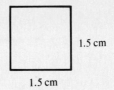

1.5 cm

1.5 cm

Solution   Start by writing the perimeter formula for a square:

$$P = 4s$$

Then evaluate the formula:

$P = 4(\textbf{1.5 cm})$ Substitute.

$P = 6$ cm   Compute.

The perimeter of the square is 6 centimeters.   ■

When the given measures have different units of measure, you must first rename to get a common unit of measure before substituting the measures into the appropriate formula.

Example 3   Find the perimeter of the following rectangle using a formula.

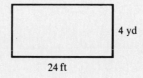

4 yd

24 ft

Solution   Start by renaming to get a common unit of measure:

$$24 \text{ ft} = 24\left(\frac{1}{3} \textbf{ yd}\right)$$ THINK: 1 yd = 3 ft or
                                1 ft = $\frac{1}{3}$ yd

$$= 8 \text{ yd}$$ See the following Note.

Then write and evaluate the perimeter formula for a rectangle:

$$P = 2(l + w)$$

$P = 2(\textbf{8 yd} + \textbf{4 yd})$ Substitute.

$P = 2(12 \text{ yd})$ Compute.

$P = 24 \text{ yd}$

The perimeter of the rectangle is 24 yards.   ■

Note: To get a common unit of measure in Example 3, you can rename 4 yd in terms of feet (instead of yards):

$$4 \text{ yd} = 4(\textbf{3 ft})$$
$$= 12 \text{ ft}$$

and then write and evaluate the perimeter formula for a rectangle:

$$P = 2(l + w)$$
$$P = 2(\mathbf{24\ ft} + \mathbf{12\ ft}) \qquad \text{Substitute.}$$
$$P = 2(36\ ft) \qquad \text{Compute.}$$
$$P = 72\ ft$$

The perimeter of the rectangle is 72 ft, or equivalently, 24 yd, because

$$24\ yd = 24 \times 3\ ft$$
$$= 72\ ft$$

6.3   FIND THE CIRCUMFERENCE

A **circle** is a closed figure in which every point is the same distance from a fixed point inside the figure. The fixed point is called the **center** of the circle.

circle

A **semicircle** is one-half of a circle.

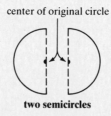

two semicircles

Any straight line segment that connects a point on a circle with its center is called a **radius** r. More than one-radius is called a **radii.**

Note: There are always an infinite number of radii that can be drawn for a given circle. All of the radii of a circle are the same length. The letter r is used to denote both a radius and also the length of that radius.

Any straight line segment that connects two points on a circle and also passes through its center is called a **diameter** d.

Note: A diameter is a line segment containing two radii.

The length of a diameter for a given circle is always twice the length of a radius.

$$\text{diameter} = 2 \times \text{radius}$$

or using symbols,

$$d = 2r$$

Example 4   Find the length of a diameter for the circle shown below:

Solution

$$d = 2r$$

$$d = 2(\textbf{3 in.}) \qquad \text{Substitute.}$$
$$d = 6 \text{ in.} \qquad \text{Compute.}$$

The length of any diameter for the circle shown above is 6 inches.   ■

The length of a radius for a given circle is always one-half the length of a diameter.

$$\text{radius} = \text{one-half of diameter}$$

or using symbols,

$$r = \frac{1}{2}d \qquad \text{or} \qquad r = \frac{d}{2}$$

Example 5   Find the length of a radius for the circle shown below:

Solution

$$r = \frac{1}{2}d \qquad \text{or} \qquad r = \frac{d}{2}$$

$$r = \frac{1}{2}(\textbf{12 cm}) \qquad\qquad r = \frac{\textbf{12 cm}}{2}$$

$$r = 6 \text{ cm} \qquad\qquad r = 6 \text{ cm}$$

The length of any radius for the circle shown above is 6 cm.   ■

The distance around a circle is called its **circumference** C. The ratio of circumference to diameter for any given circle is always **pi** (π). The number π is a nonrepeating and nonterminating decimal that is approximately equal to $3\frac{1}{7}$ or 3.14. The following are the circumference (C) formulas using diameter (d) and radius (r):

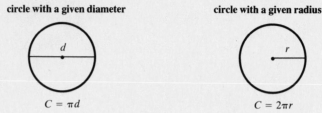

circle with a given diameter   **circle with a given radius**

$$C = \pi d \qquad\qquad\qquad C = 2\pi r$$

Note: If you memorize only $C = \pi d$, you can always derive $C = 2\pi r$ by using $d = 2r$. For example,

$$C = \pi d \longleftarrow \text{ memorized formula}$$
$$C = \pi(\mathbf{2r}) \qquad \text{Substitute } 2r \text{ for } d \text{ because } d = 2r.$$
$$C = 2\pi r \longleftarrow \text{ derived formula}$$

To find the **approximate circumference** of a circle with a given length of a diameter, you first write the circumference formula $C = \pi d$, then substitute the given diameter length for d and 3.14 for π, and then multiply.

Example 6   Find the approximate circumference of the circle shown below using $\pi \approx 3.14$:

2.8 m

Solution   Start by writing the circumference formula for a circle with a given diameter length:

$$C = \pi d$$

Then evaluate the formula:

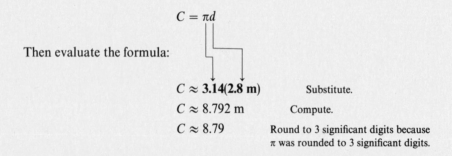

$$C \approx \mathbf{3.14(2.8\ m)} \qquad \text{Substitute.}$$
$$C \approx 8.792\ m \qquad \text{Compute.}$$
$$C \approx 8.79 \qquad \begin{array}{l}\text{Round to 3 significant digits because} \\ \pi \text{ was rounded to 3 significant digits.}\end{array}$$

The circumference of the circle is only approximately 8.79 meters

$$C \approx 8.79\ m$$

because π is only approximately 3.14

$$\pi \approx 3.14 \quad \blacksquare$$

To simplify your work and avoid introducing an approximation for π, the result of a computation involving π is often left in terms of π. For example, the circumference of the following circle in terms of π is 4π ft:

$$C = 2\pi r$$
$$C = 2\pi(\textbf{2 ft})$$
$$C = 4\pi \text{ ft} \longleftarrow \text{circumference in terms of } \pi$$

Note: The circumference of the previous circle is approximately 12.6 using $\pi \approx 3.14$:

$C = 2\pi r$ or	$C = 4\pi$ ft See previous example.
$C \approx 2(\textbf{3.14})(\textbf{2 ft})$	$C \approx 4(\textbf{3.14})$ ft
$C \approx 12.56$ ft	$C \approx 12.56$ ft
$C \approx 12.6$ ft	$C \approx 12.6$ ft (3 significant digits)

6.4 FIND THE AREA

When the surface of a closed figure is measured by finding how many squares of a given size are needed to completely cover the surface, you are finding the **area** A of that surface in **square units of measure.** The following are the common area units.

square inch (in.^{2})   **square foot** (ft^2)   **square centimeter** (cm^2)   **square meter** (m^2)

    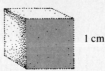

The surface of a postage stamp measures about 1 square inch (1 in.^{2}).   The surface of a $33\frac{1}{3}$ record-album cover measures about 1 square foot (1 ft^2).   Any one of the six square surfaces of a standard sugar cube measures about 1 square centimeter (1 cm^2).   The top half of a standard door measures about 1 square meter (1 m^2).

When a surface is made up of whole square units, you can count the square units to find the area.

Example 7   Find the area of the rectangle shown below by counting the number of square meters needed to completely cover the surface:

3 m

Solution   Start by dividing the given surface into square units:

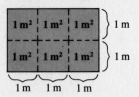

Then count the number of square units:

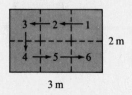

The area of the rectangle is 6 m² (square meters).   ■

When finding an area, you will usually save time if you can use a formula. The following are the common area (A) formulas using length (l), width (w), side (s), base 1 (b_1), and base 2 (b_2):

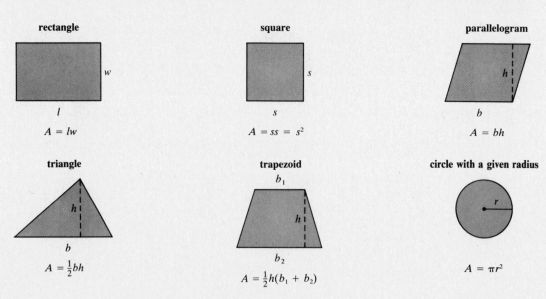

When you multiply two length measures that have the same unit of measure, such as 5 in. × 4 in., the product will always be an area measure:

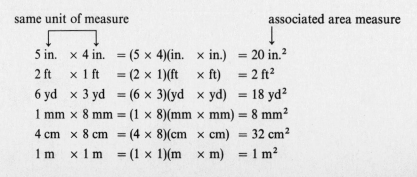

Example 8   Find the area of the rectangle from Example 1 using a formula:

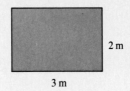

3 m

2 m

Solution   Start by writing the area formula for a rectangle:

$$A = lw$$

Then evaluate the formula:

$$A = (3\text{ m})(2\text{ m})$$ Substitute.

$$A = 6\text{ m}^2$$ Compute.

The area of the rectangle is 6 m². (The same result as found in Example 7 by counting square units.)   ■

Recall: When the given measures have different units of measure, you must first rename to get a common unit of measure before substituting the measures into the appropriate formula.

Example 9   Find the area of the triangle shown below in terms of square inches:

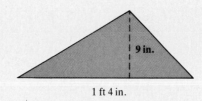

9 in.

1 ft 4 in.

Solution   Start by renaming to get a common unit of measure, inches:

$$1\text{ ft 4 in.} = 1\text{ ft} + 4\text{ in.}$$

$$= 12\text{ in.} + 4\text{ in.}$$

$$= 16\text{ in.}$$

Then write and evaluate the area formula for a triangle:

$$A = \frac{1}{2}bh$$

$$A = \frac{1}{2}(16\text{ in.})(9\text{ in.})$$ Substitute.

$$A = (8\text{ in.})(9\text{ in.})$$ Compute.

$$A = 72\text{ in.}^2$$

The area of the triangle is 72 in.² (square inches).   ■

Example 10   Find the area of the circle shown below:

Solution   Start by writing the area formula for a circle with a given radius:

$$A = \pi r^2$$

Then evaluate the formula:

$$A = \pi(\mathbf{4 \text{ cm}})^2$$
$$A = \pi(4 \text{ cm})(4 \text{ cm})$$
$$A = \pi(16 \text{ cm}^2)$$
$$A = 16\pi \text{ cm}^2 \quad \longleftarrow \quad \text{in terms of } \pi$$
$$A \approx 16(\mathbf{3.14}) \text{ cm}^2$$
$$A \approx 50.24 \text{ cm}^2 \quad \longleftarrow \quad \text{using } \pi \approx 3.14$$

The area of the circle is $16\pi \text{ cm}^2$, or approximately 50.24 cm^2 using $\pi \approx 3.14$.   ■

Note: It takes the equivalent of 50.24 cm^2 (or about 50 square centimeters) to completely cover the surface of the circle in Example 10.

To find the area of a **composite figure,** such as the one shown in Example 11, it is usually necessary to first subdivide the composite figure into two or more separate geometric figures for which you know the area formulas, such as rectangles, squares, triangles, circles, and so on.

Example 11   Find the area of the composite figure shown below:

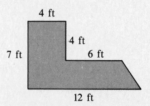

Solution   Start by subdividing the composite figure into separate geometric figures for which you know the area formulas:

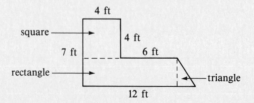

Then find the area of each subdivided figure using the appropriate geometry formula:

Square
$$A = s^2$$
$$A = (\mathbf{4 \text{ ft}})^2$$
$$A = 16 \text{ ft}^2$$

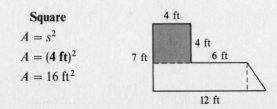

Rectangle

$A = lw$

$A = (10 \text{ ft})(3 \text{ ft})$

$A = 30 \text{ ft}^2$

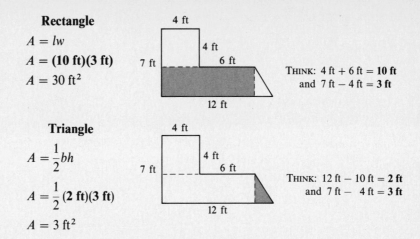

THINK: 4 ft + 6 ft = **10 ft**
and 7 ft − 4 ft = **3 ft**

Triangle

$A = \dfrac{1}{2}bh$

$A = \dfrac{1}{2}(2 \text{ ft})(3 \text{ ft})$

$A = 3 \text{ ft}^2$

THINK: 12 ft − 10 ft = **2 ft**
and 7 ft − 4 ft = **3 ft**

And then add the subdivided areas together to get the total area of the composite figure:

Composite Figure

$A = 16 \text{ ft}^2 + 30 \text{ ft}^2 + 3 \text{ ft}^2$

$A = 49 \text{ ft}^2$ ⟵——— total area

6.5   FIND THE VOLUME

A three-dimensional figure has length, width, and height. For example, a **rectangular prism** or **box** is a three-dimensional figure shaped like a standard cardboard box.

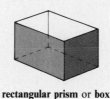

rectangular prism or box

Note: A rectangular prism, or box, has six rectangularly shaped sides.

A **cube** is a box in which each side is square.

cube

A **cylinder** is a three-dimensional figure shaped like a standard soup can.

cylinder

A **sphere** is a three-dimensional figure shaped like a round ball.

sphere

A **hemisphere** is one-half of a sphere.

hemisphere

When the space occupied by a three-dimensional object is measured by finding how many cubes of a given size are needed to completely fill that space, you are finding the **volume** V of that object in **cubic units of measure.** The following are the most common volume units.

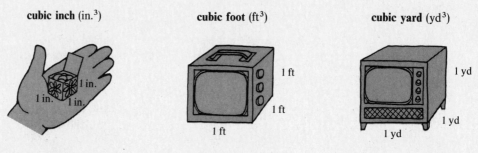

cubic inch (in.^{3})	cubic foot (ft^3)	cubic yard (yd^3)
The space occupied by this small pill box is about 1 cubic inch (1 in.^{3}).	The space occupied by a small portable TV set is about 1 cubic foot (1 ft^3).	The space occupied by a standard floor-model TV set is about 1 cubic yard (1 yd^3).

cubic centimeter (cm^3)   cubic meter (m^3)

This 1 cm by 1 cm by 1 cm box occupies 1 cubic centimeter (1 cm^3) of space and holds 1 milliliter (1 mL) of liquid.

A cube that is 1m by 1 m by 1 m will occupy 1 cubic meter (1 m^3) of space and will hold 1 kiloliter (1 kL) of liquid.

Note: Another common abbreviation for cubic centimeter is **cc.** Both cc and cm^3 mean the same thing and are used interchangably.

When the space to be filled is made up of whole cubic units, you can count the cubic units to find the volume.

Example 12   Find the volume of the rectangular box shown below by counting the number of cubic units needed to completely fill it.

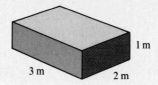

Solution   Start by dividing the given object into cubic units:

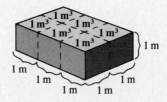

Then count the number of cubic units:

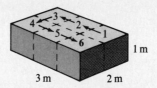

The volume of the rectangular box is 6 m³ (cubic meters).   ■

When finding a volume, you usually will save time if you can use a formula. The following are the common volume (V) formulas using area of the base (A), edge (e), length (l), width (w), height (h), and radius (r).

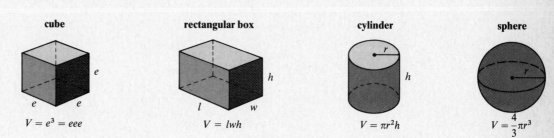

cube   **rectangular box**   **cylinder**   **sphere**

$V = e^3 = eee$ $V = lwh$ $V = \pi r^2 h$ $V = \dfrac{4}{3}\pi r^3$

When you multiply three length measures that have the same unit of measure, such as 1 in. × 4 in. × 1 in., the product will always be a volume measure:

same unit of measure   associated volume measure

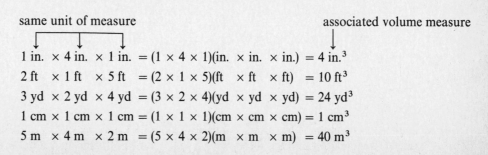

$1 \text{ in.} \times 4 \text{ in.} \times 1 \text{ in.} = (1 \times 4 \times 1)(\text{in.} \times \text{in.} \times \text{in.}) = 4 \text{ in.}^3$

$2 \text{ ft} \times 1 \text{ ft} \times 5 \text{ ft} = (2 \times 1 \times 5)(\text{ft} \times \text{ft} \times \text{ft}) = 10 \text{ ft}^3$

$3 \text{ yd} \times 2 \text{ yd} \times 4 \text{ yd} = (3 \times 2 \times 4)(\text{yd} \times \text{yd} \times \text{yd}) = 24 \text{ yd}^3$

$1 \text{ cm} \times 1 \text{ cm} \times 1 \text{ cm} = (1 \times 1 \times 1)(\text{cm} \times \text{cm} \times \text{cm}) = 1 \text{ cm}^3$

$5 \text{ m} \times 4 \text{ m} \times 2 \text{ m} = (5 \times 4 \times 2)(\text{m} \times \text{m} \times \text{m}) = 40 \text{ m}^3$

Example 13   Find the volume of the rectangular box from Example 12 using a formula:

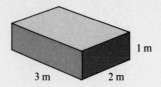

3 m   2 m   1 m

Solution   Start by writing the volume formula for a rectangular box:

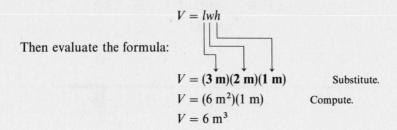

$$V = lwh$$

Then evaluate the formula:

$$V = (\mathbf{3\ m})(\mathbf{2\ m})(\mathbf{1\ m}) \qquad \text{Substitute.}$$
$$V = (6\ m^2)(1\ m) \qquad \text{Compute.}$$
$$V = 6\ m^3$$

The volume of the rectangular box is 6 m³. (The same result as found in Example 12 by counting cubic units.)   ■

Example 14   Find the volume of the cylinder shown below:

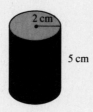

2 cm

5 cm

Solution   Start by writing the volume formula for a cylinder:

$$V = \pi r^2 h$$

Then evaluate the formula:

$$V = \pi(\mathbf{2\ cm})^2(\mathbf{5\ cm})$$
$$V = \pi(4\ cm^2)(5\ cm) \qquad \text{THINK: } (2\ cm)^2 = (2\ cm)(2\ cm)$$
$$V = \pi(20\ cm^3)$$
$$V = 20\pi\ cm^3 \quad \longleftarrow \text{ in terms of } \pi$$
$$V \approx 20(\mathbf{3.14})\ cm^3$$
$$V \approx 62.8\ cm^3 \quad \longleftarrow \text{ using } \pi \approx 3.14$$

The volume of the cylinder is exactly 20π cm³, or approximately 62.8 cm³ using π ≈ 3.14.   ■

Note: It takes the equivalent of 62.8 cm³ (or about 63 cubic centimeters) to completely fill the cylinder in Example 13.

6.6   FIND THE SURFACE AREA

The total area of all the surfaces of a 3-dimensional geometric object is called its **surface area.** When the surface to be measured is made up of whole square units, you can often count the square units to find the surface area.

Example 15   Find the surface area of the rectangular box shown below by counting the number of square units needed to completely cover each side.

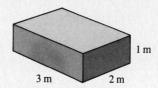

Solution   Start by dividing the given object into square units:

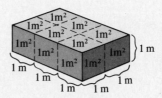

Then count the number of square units:

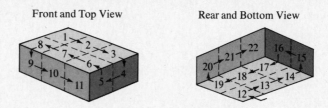

Front and Top View   Rear and Bottom View

The surface area of the rectangular box is 22 m² (square meters).   ■

When finding a surface area, you will usually save time if you can use a formula. The following are the common surface area (*SA*) formulas using edge (*e*), length (*l*) width (*w*), height (*h*), and radius (*r*).

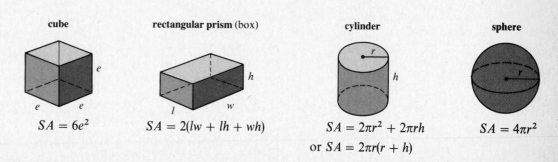

| cube | rectangular prism (box) | cylinder | sphere |

$$SA = 6e^2 \qquad SA = 2(lw + lh + wh) \qquad SA = 2\pi r^2 + 2\pi rh \qquad SA = 4\pi r^2$$

$$\text{or } SA = 2\pi r(r + h)$$

Example 16   Find the surface area of the rectangular box shown below using a formula:

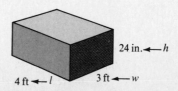

Solution   Start by writing the surface area formula for a rectangular box:

$$SA = 2(lw + lh + wh)$$

Then evaluate the formula: 24 in. = 2 ft

$$SA = 2[(4 \text{ ft})(3 \text{ ft}) + (4 \text{ ft})(2 \text{ ft}) + (3 \text{ ft})(2 \text{ ft})]$$
$$SA = 2[12 \text{ ft}^2 + 8 \text{ ft}^2 + 6 \text{ ft}^2]$$
$$SA = 2[26 \text{ ft}^2]$$
$$SA = 52 \text{ ft}^2$$

The surface area of the rectangular box is 52 ft². ∎

CHAPTER 6 PRACTICE

For exercises 1–12, match each figure on the left with one and only one name on the right. Do this in such a way that no two figures have the same name. [See pages 84–86, 88, and 95–96.]

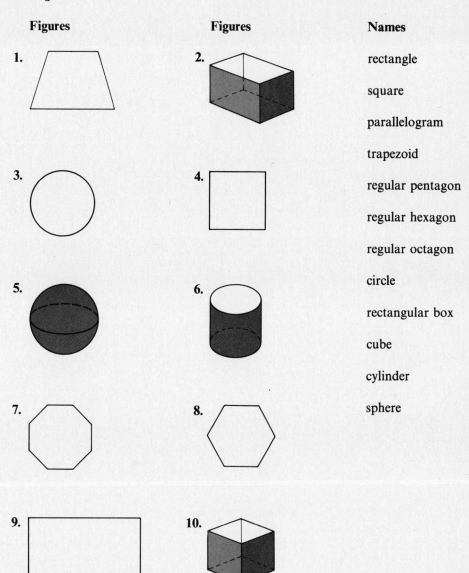

Figures	**Figures**	**Names**
1.	2.	rectangle
		square
		parallelogram
		trapezoid
3.	4.	regular pentagon
		regular hexagon
		regular octagon
5.	6.	circle
		rectangular box
		cube
7.	8.	cylinder
		sphere
9.	10.	
11.	12.	

Copyright © 1990 by Harcourt Brace Jovanovich, Inc. All rights reserved.

For exercises 13–24, find the perimeter of each polygon.

[See Example 1.]

13.

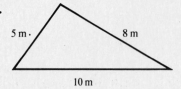

5 m ·   8 m

10 m

14.

6 ft   5 ft

3 ft

15.

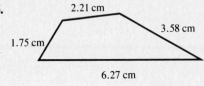

2.21 cm

3.58 cm

1.75 cm

6.27 cm

16.

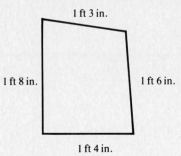

1 ft 3 in.

1 ft 8 in.   1 ft 6 in.

1 ft 4 in.

[See Example 2.]

17.

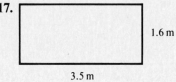

1.6 m

3.5 m

18.

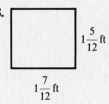

$1\frac{5}{12}$ ft

$1\frac{7}{12}$ ft

19.

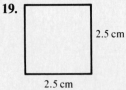

2.5 cm

2.5 cm

20.

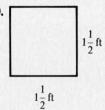

$1\frac{1}{2}$ ft

$1\frac{1}{2}$ ft

[See Example 3.]

21.

175 cm

22.

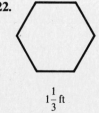

$1\frac{1}{3}$ ft

Copyright © 1990 by Harcourt Brace Jovanovich, Inc. All rights reserved.

23.

500 cm

6 m

24.

8 in.

8 in.

For each exercise 25–30, find the length of a diameter of a circle given that the length of a radius for the same circle is:

[See Example 4.]

25. 4 ft   **26.** 6 in.   **27.** 1.4 cm

28. 0.8 m   **29.** 2.5 mm   **30.** $3\frac{1}{4}$ yd

For each exercise 31–36, find the length of a radius of a circle given that the length of a diameter for the same circle is:

[See Example 5.]

31. 1 yd   **32.** 2 in.   **33.** 1.5 mm

34. 0.6 cm   **35.** 3.25 m   **36.** $4\frac{2}{3}$ ft

For exercises 37–42, find the circumference of each circle **a.** in terms of π **b.** using $\pi \approx 3.14$. [See Example 6.]

37.
3 cm

38.
9 m

39.
21 ft

40.
28 yd

41.
8 mm

42.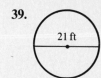
7 in.

For exercises 43–46, find the area of each surface by counting square units. [See Example 7.]

43.

2 ft

2 ft

44.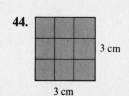

3 cm

3 cm

Copyright © 1990 by Harcourt Brace Jovanovich, Inc. All rights reserved.

45.

3 in.

2 in.

46.

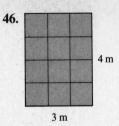

4 m

3 m

For exercises 47–50, find the area of each surface using the appropriate formula. [See Example 8.]

47.

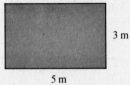

3 m

5 m

48.

5.6 m

340 cm

[See Example 9.]

49.

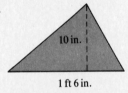

10 in.

1 ft 6 in.

50.

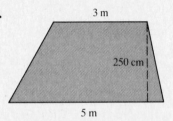

3 m

250 cm

5 m

For exercises 51–54, find the area of each circle **a.** in terms of π **b.** using $\pi \approx 3.14$. [See Example 10.]

51.

3 mm

52.

$3\frac{1}{2}$ in.

53.

5 m

54.

$1\frac{3}{4}$ ft

For exercises 55–62, find the area of each composite figure. [See Example 11.]

55.

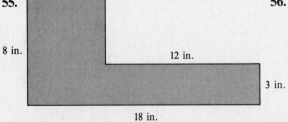

8 in.

12 in.

3 in.

18 in.

56.

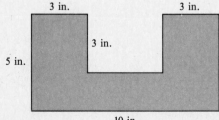

3 in.   3 in.

3 in.

5 in.

10 in.

Copyright © 1990 by Harcourt Brace Jovanovich, Inc. All rights reserved.

57.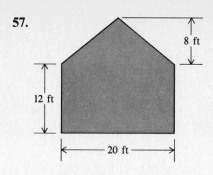

58.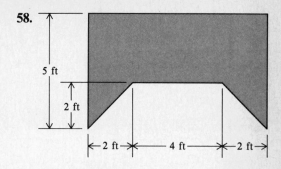

59.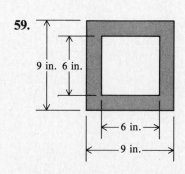

60.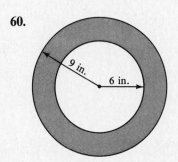

61.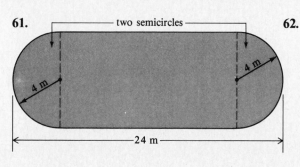

62.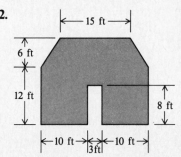

For exercises 63–66, find the volume of each rectangular box by counting cubic units. [See Example 12.]

63.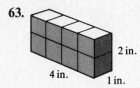

64.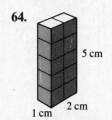

65.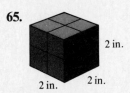

66.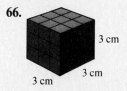

Copyright © 1990 by Harcourt Brace Jovanovich, Inc. All rights reserved.

For exercises 67–70, find the **a.** volume and **b.** surface area of each object using the appropriate formula. [See Example 13.]

67.

4 in.

8 in.

6 in.

68.

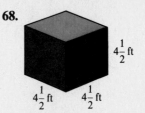

$4\frac{1}{2}$ ft

$4\frac{1}{2}$ ft $4\frac{1}{2}$ ft

69.

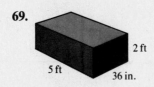

2 ft

5 ft   36 in.

70.

5 yd

3 yd

2 ft

For exercises 71 and 72, find the **a.** volume and **b.** surface area in terms of π and using $\pi \approx 3.14$. [See Examples 14 and 16.]

71.

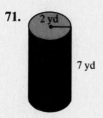

2 yd

7 yd

72.

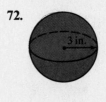

3 in.

Copyright © 1990 by Harcourt Brace Jovanovich, Inc. All rights reserved.

PROPERTIES OF POLYGONS, CIRCLES, AND SOLIDS

In this chapter you will solve geometry problems that involve:

- Properties of common polygons.
- Properties of circles.
- Properties of common 3-dimensional figures.

| 7.1 | **USE PROPERTIES OF COMMON POLYGONS** |

Recall: The sum of the interior angles of any triangle is 180°.

To find the sum of the interior angles of any given polygon, such as a quadrilateral (4 sides), a pentagon (5 sides), a hexagon (6 sides), or an octagon (8 sides), you can always subdivide the given polygon into separate triangles and then make use of the fact that each of the separate triangles contains 180°.

Example 1   Find the sum of the interior angles of any quadrilateral:

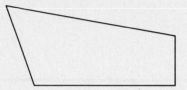

Solution   Start by subdividing the quadrilateral into separate triangles in such a way that the vertex of each separate triangle is also a vertex of the given figure:

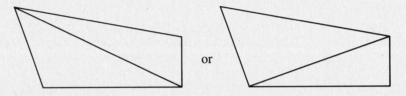

Then make use of the fact that each separate triangle contains 180° by multiplying 180° by the number of separate triangles formed:

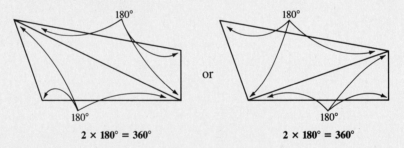

And so, the sum of the interior angles of any quadrilateral is 360°:

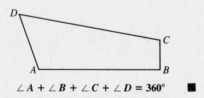

$$\angle A + \angle B + \angle C + \angle D = 360°$$ ∎

Note: In Example 1, the given quadrilateral could have been any four-sided polygon. In particular, the sum of the interior angles of a rectangle, square, parallelogram, and trapezoid are all 360°:

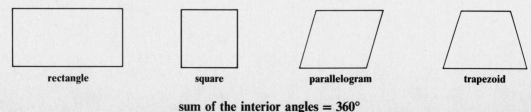

rectangle   square   parallelogram   trapezoid

sum of the interior angles = 360°

CAUTION   To find the sum of the interior angles of a polygon by subdividing it into separate triangles, the vertex of each separate triangle **must** also be a vertex of the given polygon. For example, the following quadrilateral *ABCD* has been subdivided into separate triangles incorrectly, because the point *E* is not a vertex of the original quadrilateral *ABCD*:

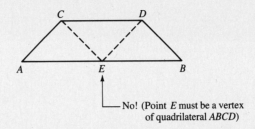

└─ No! (Point *E* must be a vertex
of quadrilateral *ABCD*)

Note: The sum of the interior angles of quadrilateral *ABCD* is 360°, not 540°, as the incorrect subdivision would suggest.

Verify that the sum of the interior angles of any pentagon is 540°, not 360°, as in a quadrilateral, using the method shown in Example 1.

A very common type of geometry problem, is one in which you are given a polygon, and one or more properties of the given polygon, and then asked to find the perimeter or area of the polygon using the given properties.

Example 2   Given the following regular hexagon and the fact that the perimeter of each of the six equilateral triangles is $\sqrt{5}$ units, find the perimeter of the regular hexagon.

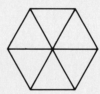

Solution   The perimeter formula for an equilateral triangle is

$$P = 3s$$

and so,

given $\longrightarrow$ $\sqrt{5} = 3s$

$$s = \frac{\sqrt{5}}{3} \text{ (units)} \longleftarrow \text{ length of each side of the regular pentagon}$$

The perimeter formula for a regular hexagon is

$$P = 6s$$

and so,

$$P = 6\left(\frac{\sqrt{5}}{3}\right) \qquad \text{Substitute.}$$

$$P = 2\sqrt{5} \text{ (units)} \qquad \text{Simplify.}$$

The perimeter of the given regular hexagon is $2\sqrt{5}$ units (≈ 4.472 units).   ∎

Recall: To solve a problem using a formula, given specific values for all the variables in that formula except one, you substitute each of the known values for the corresponding variables and then solve the resulting equation in one variable.

Example 3   If the area of a rectangle is 180 feet and its length is 12 feet, then the width and perimeter of the rectangle are 15 feet and 54 feet, respectively.   ■

Verify Example 3.

Note: Given the perimeter of a square, you can always find its area. And, conversely, given the area of a square, you can always find its perimeter. For example, if the perimeter of a square is 36 cm, then its area is 81 cm². And, if the area of a square is 36 cm², then its perimeter is 24 cm.

Verify both examples in the Note.

Recall: To solve a problem using a formula when there are two different variables in the formula for which you do not have specific values, you must be able to represent one of the two variables in terms of the other.

Example 4   If the perimeter of a rectangle is 64 feet and its length is 10 feet longer than its width, then the length, width, and area of the rectangle are 21 feet, 11 feet, and 231 square feet, respectively.   ■

Verify Example 4.

7.2   USE PROPERTIES OF CIRCLES

Given the circumference of a circle, you can always find its area. And, conversely, given the area of a circle, you can always find its circumference.

Example 5   If the circumference of a circle is 16π cm, then its area is 64π cm². And, if the area of a circle is 16π cm², then its circumference is 8π cm.   ■

Verify Example 5.

The following properties of circles are very useful in solving common geometry problems.

> **PROPERTIES OF CIRCLES**
> **1.** The ratio of the circumferences of any two circles equals the ratio of the lengths of their *radii* (radii is plural for radius).
> **2.** The ratios of the areas of any two circles equals the ratio of the squares of the lengths of their radii.

For example, if the circumference of the following two circles are represented by C and C', respectively, and the areas are represented by A and A', respectively,

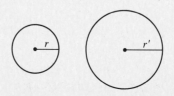

then

$$\frac{C}{C'} = \frac{r}{r'}$$

and

$$\frac{A}{A'} = \frac{r^2}{r'^2}$$

Example 6 Given two circles with radii of 12 cm and 18 cm, respectively, what is the ratio of their circumferences?

Solution Start by substituting the given radii in the proportion relating circumference and radius:

$$\frac{C}{C'} = \frac{r}{r'}$$

$$\frac{C}{C'} = \frac{12}{18} \quad \begin{array}{l} \longleftarrow \text{ radius of smaller circle} \\ \longleftarrow \text{ radius of larger circle} \end{array}$$

$$\frac{C}{C'} = \frac{2}{3}$$

The ratio of the smaller circle's circumference to the larger circle's circumference is

$$2 \text{ to } 3 \quad \text{ or } \quad 2{:}3 \quad \text{ or } \quad \frac{2}{3} \quad \blacksquare$$

Note: The ratio of the larger circle's circumference to the smaller circle's circumference is

$$3 \text{ to } 2 \quad \text{ or } \quad 3{:}2 \quad \text{ or } \quad \frac{3}{2}$$

Example 7 Given two circles whose radii are in a ratio of $\frac{3}{4}$, find the area of the smaller circle if the area of the larger circle is 157 cm^2.

Solution Because the numerator of $\frac{3}{4}$ is smaller than the denominator, the numerator represents the smaller circle and the denominator represents the larger circle. Start by substituting the given values in the direct proportion relating the areas and the squares of the radii:

$$\frac{A}{A'} = \frac{r^2}{r'^2}$$

$$\begin{array}{l} \text{smaller circle} \longrightarrow A \\ \text{larger circle} \longrightarrow \overline{157 \text{ cm}^2} \end{array} = \frac{3^2}{4^2} \begin{array}{l} \longleftarrow \text{ smaller circle} \\ \longleftarrow \text{ larger circle} \end{array}$$

And then solve for A:

$$16A = 9(157 \text{ cm}^2)$$

$$A = \frac{1413}{16} \text{ cm}^2$$

$$A = 88.3125 \text{ cm}^2$$

The area of the smaller circle is 88.3125 cm^2. $\blacksquare$

Example 8   If the circumference of circle A is four times the circumference of circle B, and the radius of circle B is 5 cm, what is the radius of circle A?

$$\text{Let } \quad r = \text{the radius of circle } A \text{ (the larger circle)}$$
$$\text{and } \quad r' = \text{the radius of circle } B \text{ (the smaller circle)}.$$

Then substitute the given values in the direct proportion relating circumference and radius:

$$\frac{C}{C'} = \frac{r}{r'}$$

larger circle $\longrightarrow$ $\dfrac{\mathbf{4}}{\mathbf{1}} = \dfrac{r}{\mathbf{5\ cm}}$ $\longleftarrow$ larger circle
smaller circle $\longrightarrow$ $\qquad\qquad$ $\longleftarrow$ smaller circle

And then solve for r:

$$lr = 4(5\text{ cm})$$
$$r = 20\text{ cm}$$

The radius of circle A is 20 cm.   ■

The following problem involves a polygon and a circle.

Example 9   Given the following figure, find the circumference of the inscribed circle if the area of the square is 36 m².

Solution   Start by finding the length of each side of the square:

$$A = s^2 \longleftarrow \text{ area formula for a square}$$
$$\mathbf{36} = s^2$$
$$s = 6\text{ (m)} \longleftarrow \text{ length of each side}$$

And so, the diameter of the inscribed circle is 6 m because the length of a diameter of the inscribed circle and the length of a side of the square are the same:

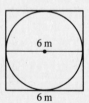

6 m

6 m

And so, the circumference of the inscribed circle is 6π m (≈ 18.84 m) because

$$C = \pi d$$
$$C = \pi(\mathbf{6\ m})$$
$$C = 6\pi\text{ m} \quad ■$$

Note 1: The area of the inscribed circle in Example 9 is 9π m^2 (≈ 28.26 m^2) because

$$A = \pi r^2$$
$$A = \pi(\mathbf{3\ m})^2$$
$$A = 9\pi\ \text{m}^2$$

Note 2: The area of the part of the square that is outside the inscribed circle in Example 9 is $(36 - 9\pi)$ m^2 (≈ 7.74 m^2) because

$$\text{area of square} - \text{area of circle} = \mathbf{36\ m^2 - 9\pi\ m^2}$$
$$= (36 - 9\pi)\ \text{m}^2$$

7.3	USE PROPERTIES OF SOLIDS

A common type of 3-dimensional geometry problem is one in which the volume of a cube is given, and you are asked to find its dimensions.

Example 10   Find the dimension of a cube whose volume is 27 m^3.

Solution   Start by writing the volume formula for a cube:

$$V = s^3$$

Then substitute 27 cm^3 for V and solve for s:

$$\mathbf{27\ m^3} = s^3$$
$$s = 3\ \text{m}$$

Check: $(3\ \text{m})(3\ \text{m})(3\ \text{m}) = 27\ \text{m}^3 \longleftarrow s = 3$ m checks   ■

The dimensions of the cube are 3 m × 3 m × 3 m. Read 3 m × 3 m × 3 m as "3 meters by 3 meters by 3 meters."

Note: Given the volume of a cube, you can always find its dimensions.

Another common type of geometry problem involving two 3-dimensional figures, is one in which you are given the relationship between two or more corresponding dimensions, and you are asked to find the relationship between the two volumes.

Example 11   If the length of each side of a cube is doubled, by how much is the volume increased?

Solution   Start by drawing and labeling the two cubes:

Let x = the length of each side of the smaller cube

then $2x$ = the length of each side of the larger cube.

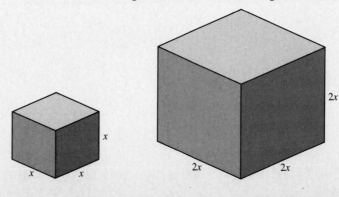

Then compute and compare the two volumes:

$$V = s^3$$
$$V = (\boldsymbol{x})^3$$
$$V = x^3 \longleftarrow \text{ volume of smaller cube}$$
$$V = s^3$$
$$V = (\boldsymbol{2x})^3$$
$$V = 8x^3 \longleftarrow \text{ volume of larger cube}$$

And so, the volume is increased 8 times when the length of each side of a cube is doubled.   ∎

CHAPTER 7 PRACTICE

For each exercise 1–12, write "true" or "false."

1. The sum of the interior angles of a square is 360°.

2. The sum of the interior angles of a rectangle is 360°.

3. The sum of the interior angles of a parallelogram is 360°.

4. The sum of the interior angles of a pentagon is 360°.

5. The ratio of the circumference of any two circles equals the ratio of the lengths of their radii.

6. The ratios of the areas of any two circles equals the ratio of the lengths of their radii.

7. The length of a diameter of a circle inscribed in a square is the same as the length of any side of the square.

8. The length of a diameter of a circle circumscribed around a square (intersecting the four vertices of the square) is the same as the diagonal of the square.

9. Given the perimeter of a square, you can always find its area.

10. Given the area of a rectangle, you can always find its perimeter.

11. Given the area of a circle, you can always find its circumference.

12. Given the volume of a cube, you can always find its dimensions.

For each exercise 13–18, find the sum of the interior angles of the given polygon. [See Example 1.]

13. triangle

14. quadrilateral

15. pentagon

16. hexagon

17. octagon

18. trapezoid

19. Find the perimeter of the following regular hexagon if the perimeter of each of the six equilateral triangles is 2 units.

20. Find the perimeter of the following regular hexagon if the perimeter of each of the six equilateral triangles is $\sqrt{3}$ units.

Copyright © 1990 by Harcourt Brace Jovanovich, Inc. All rights reserved.

For each exercise 21–26, fill in the blanks. [See Example 3.]

Figure	Length	Width	Perimeter	Area
21. rectangle	3 ft	6 ft	_____	_____
22. rectangle	_____	15 m	_____	240 m²
23. rectangle	16 in	_____	92 in.	_____
24. square	6 yd	_____	_____	_____
25. square	_____	_____	_____	64 cm²
26. square	_____	_____	64 cm	_____

[See Example 4.]

27. The perimeter of a rectangle is 10 feet. The width is 3 feet less than the length. What is its area?

28. The perimeter of a rectangle is 110 feet. The length is 5 feet shorter than twice the width. Find the area of the rectangle.

For each exercise 29–34, fill in the blanks. [See Example 5.]

Figure	Radius	Diameter	Circumference	Area
29. circle	5 in.	_____	_____	_____
30. circle	_____	4 ft	_____	_____
31. circle	_____	_____	36π m	_____
32. circle	_____	_____	_____	36π m²
33. circle	_____	_____	25.12 cm (use $\pi \approx 3.14$)	_____
34. circle	_____	_____	_____	28.26 yd² (use $\pi \approx 3.14$)

Copyright © 1990 by Harcourt Brace Jovanovich, Inc. All rights reserved.

[See Example 6.]

35. Given that two circles have radii of 12 m and 15 m respectively, what is the ratio of the smaller circle's circumference to the larger circle's circumference?

36. In problem 35, what is the ratio of the larger circle's area to the smaller circle's area?

[See Example 7.]

37. Given two circles whose radii are in a ratio of $\frac{2}{3}$, find the area of the smaller circle if the area of the larger circle is 48 m^2.

38. Given two circles whose radii are in a ratio of $\frac{1}{4}$, find the circumference of the larger circle if the circumference of the smaller circle is 7.56 cm.

[See Example 8.]

39. If the circumference of circle A is three times the circumference of circle B, and the radius of circle B is 3 cm, what is the radius of circle A?

40. If the area of circle A is one-half the area of circle B, and the radius of circle A is 8 m, what is the radius of circle B?

[See Example 9.]

41. If the area of the square in the following figure is 8 cm^2, find the **a.** circumference, **b.** area of the inscribed circle, and **c.** the area of the square that is outside the circle.

42. If the area of the circle in the figure below is 4π m^2, find the **a.** perimeter, **b.** area of the inscribed square, and **c.** area of the circle that is outside the square.

For each exercise 43–52, fill in the blank(s). [See Example 10.]

Figure	Length	Width	Height	Volume
43. cube	2 ft	_____	_____	_____
44. cube	_____	_____	_____	64 cm^3
45. box	2 m	3 m	4 m	_____
46. box	_____	5 mm	6 mm	120 mm^3

Copyright © 1990 by Harcourt Brace Jovanovich, Inc. All rights reserved.

	Radius	Diameter	Height	Volume
47. sphere	$2\frac{1}{2}$ ft	_____	not applicable	_____
48. sphere	_____	4 yd	not applicable	_____
49. sphere	_____	_____	not applicable	36π m³
50. cylinder	2 cm	_____	4 cm	_____
51. cylinder	_____	8 in.	_____	112π in.³
52. cylinder	_____	_____	4 ft	314 ft³ (use $\pi \approx 3.14$)

[See Example 11.]

53. If the length of each side of a cube is tripled, by how many times is the volume increased?

54. If the radius of a sphere is doubled, by what factor is the volume increased?

Mixed Practice

55. Find the area of a circle whose circumference is 24π.

56. Find the circumference of a circle whose area is 62.8, using $\pi \approx 3.14$.

57. How many cubic feet of water are in a one-half full-cylindrical water tank with radius 2 feet and height 8 feet?

58. Find the radius of a circle whose area is numerically equal to its circumference.

59. What is the perimeter of the figure shown below, given that the figure is composed of a rectangle and two semicircles?

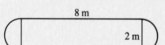

60. What is the area of the figure shown below, given that the figure is composed of a square and two semicircles?

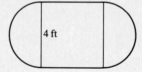

61. Find the area of the small circle shown below, in terms of π.

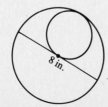

62. Find the perimeter of the square shown below.

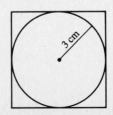

Copyright © 1990 by Harcourt Brace Jovanovich, Inc. All rights reserved.

63. Find $\angle X$ in the figure below given:

$$XZ \cong ZY \cong YP \cong PZ$$

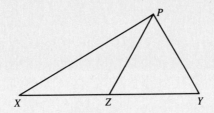

64. If you double the length and width of a rectangular box and leave the height the same, by how much is the volume increased?

Copyright © 1990 by Harcourt Brace Jovanovich, Inc. All rights reserved.

SELECTED ANSWERS

CHAPTER 1 PRACTICE

1.1 PRACTICE, p. 3
1. 2:15   **3.** Spikers: W-1, L-3; Setters: W-3, L-1; Servers: W-2, L-2

1.2 PRACTICE, p. 7
1. 6   **3.** 10   **5.** 14   **7.** 5   **9.** 8   **11.** 06   **13.** B

1.3 PRACTICE, p. 11
1. You are my sister-in-law.   **3.** You are my aunt.   **5.** You are my cousin.   **7.** You are my son-in-law.
9. You are my father or You are my stepfather.   **11.** You are my grandmother.

1.4 PRACTICE, pp. 15–16
1. No
3. The box marked "R" contains blue marbles, the box marked "B" contains both red and blue marbles, and the box marked "RB" contains red marbles.
5. Alan   **7.** Butler

CHAPTER 2 PRACTICE

1. 2 in.   **3.** $1\frac{3}{4}$ in.   **5.** $1\frac{13}{16}$ in.   **7.** 46 mm   **9.** 5.08 cm   **11.** 1.8288 m   **13.** 91.44 m   **15.** 59.055 in.   **17.** 16.4 ft   **19.** 6.214 mi
21. 40°   **23.** 130°   **25.** 25°   **27.** 140°   **29.** 115°   **31.** 180°   **33.** 36.25°   **35.** 82.75°   **37.** 105.33°   **39.** 42.01°   **41.** 6.30°
43. 53.74°   **45.** 17°30′   **47.** 46°45′   **49.** 103°21′   **51.** 39°6′36″   **53.** 130°30′36″   **55.** 82°35′24″   **57.** equilateral triangle
59. straight angle   **61.** right triangle   **63.** obtuse angle   **65.** isosceles triangle   **67.** obtuse triangle

CHAPTER 3 PRACTICE

1. True   **3.** True   **5.** True   **7.** True   **9.** True   **11.** True   **13.** True   **15.** True   **17.** True   **19.** l_1 and l_2 **21.** l_3 **23.** None
25. $\angle p$ and $\angle r$, $\angle q$ and $\angle s$, $\angle t$ and $\angle v$, $\angle u$ and $\angle w$ **27.** $\angle p$ and $\angle t$, $\angle q$ and $\angle u$, $\angle r$ and $\angle v$, $\angle s$ and $\angle w$

29. $\angle p = 45°$, $\angle q = 135°$, $\angle r = 45°$, $\angle s = 135°$, $\angle t = 45°$, $\angle u = 135°$, $\angle v = 45°$, $\angle w = 135°$ **31.** $\angle a = 115°$ **33.** $\angle a = 55°$ **35.** $\frac{4}{3}$
37. Yes   **39.** $-\frac{5}{6}$ **41.** $\angle a = 35°$ **43.** $\angle d = 45°$ **45.** $\angle c = 60°$ **47.** $\angle g = 110°$ **49.** $\angle a = 25°$ **51.** $\angle a = 60°$ **53.** $\angle c = 10°$
55. $\frac{4}{5}$ **57.** $-\frac{5}{4}$ **59.** $-\frac{2}{5}$

CHAPTER 4 PRACTICE

1. True   **3.** True   **5.** True   **7.** True   **9.** False   **11.** True   **13.** $F = 15°$ **15.** $\angle A = 40°$, $\angle C = 100°$
17. $\angle A = 60°$, $\angle B = 60°$, $\angle C = 60°$ **19.** 13 cm   **21.** Yes   **23.** hypotenuse $= \sqrt{41}$ ft ≈ 6.4 ft (4 ft, 5 ft, $\sqrt{41}$ ft)   **25.** $\angle C = 63°$
27. $\angle B = 50°$, $\angle C = 80°$ **29.** $\angle A = 60°$, $\angle B = 60°$, $\angle C = 60°$ **31.** $\angle B = 60°$, $\angle C = 90°$ **33.** $\angle B = 57°$ **35.** $\angle A = 52°$
37. $c = 15$ in.   **39.** $b = 10$ yd   **41.** $a = 9$ cm   **43.** $c = \sqrt{3}$ in.   **45.** $c = 5$ ft   **47.** $c = 6.5$ cm   **49.** 65 ft   **51.** 100 m   **53.** 410 km
55. $\sqrt{13}$ ft or $\sqrt{5}$ ft

CHAPTER 5 PRACTICE

1. True   **3.** True   **5.** True   **7.** False   **9.** True   **11.** False   **13. a.** Yes   **b.** They satisfy SSS.
15. a. No   **b.** BC must be congruent to YZ to satisfy ASA or AB must be congruent to XY to satisfy AAS.
17. a. No   **b.** similar, not congruent
19. a. Yes   **b.** Two triangles are similar if only two pairs of corresponding angles are congruent.
21. a. Yes   **b.** Two triangles are similar if there are two corresponding angles (or common angles) formed by a transversal intersecting two parallel lines, and the corresponding sides opposite those angles are parallel.
23. a. Yes   **b.** Two triangles are similar if there are two vertical angles and the pair of corresponding sides opposite the vertical angles are parallel.
25. $AB = 10.8$ m   **27.** $YZ = 24$ in., $XZ = 26\frac{2}{3}$ in.   **29.** $DE = 4.8$ km   **31.** $DE = 5\frac{1}{3}$ mi, $CE = 8$ mi, $AD = 25$ mi, $BE = 20$ mi
33. $AC' = 5\frac{1}{3}$ m   **35.** $AC' = 22.5$ cm   **37.** $AC = 3$ cm, $AB = 3\sqrt{2}$ cm   **39.** $AC = \dfrac{7}{\sqrt{2}}$ or $\dfrac{7\sqrt{2}}{2}$ km, $BC = \dfrac{7}{\sqrt{2}}$ or $\dfrac{7\sqrt{2}}{2}$ km
41. $YZ = 5$ mi, $XY = 5\sqrt{2}$ mi   **43.** $XZ = \dfrac{2}{\sqrt{2}}$ or $\sqrt{2}$ ft, $YZ = \dfrac{2}{\sqrt{2}}$ or $\sqrt{2}$ ft   **45.** $AC = 2\sqrt{2}$ m   **47.** $CD = \dfrac{20}{\sqrt{2}}$ or $10\sqrt{2}$ mi
49. $BC = 4\sqrt{3}$ m, $AB = 8$ m   **51.** $AC = \dfrac{10}{\sqrt{3}}$ or $\dfrac{10\sqrt{3}}{3}$ ft, $AB = \dfrac{20}{\sqrt{3}}$ or $\dfrac{20\sqrt{3}}{3}$ ft   **53.** $AC = 10$ km, $BC = 10\sqrt{3}$ km
55. $BD = 4\sqrt{2}$ cm, $AC = 4\sqrt{3}$ cm, $AB = 8$ cm   **57.** $AC = \dfrac{9\sqrt{3}}{2}$ in., $DC = 4\frac{1}{2}$ in., $BD = \dfrac{9\sqrt{2}}{2}$ in.
59. $AB = 4$ km, $BD = 2\sqrt{2}$ km, $AD = (2 + 2\sqrt{3})$ km   **61.** $AB = 1$ **63.** $\frac{4}{5}$ **65.** 36 ft   **67.** 860,000 mi

CHAPTER 6 PRACTICE

1. trapezoid   **3.** circle   **5.** sphere   **7.** regular octagon   **9.** rectangle   **11.** regular pentagon   **13.** 23 m   **15.** 13.81 cm   **17.** 10.2 m
19. 10 cm   **21.** 1400 cm   **23.** 22 m or 2200 cm   **25.** 8 ft   **27.** 2.8 cm   **29.** 5 mm   **31.** $\frac{1}{2}$ yd   **33.** 0.75 mm   **35.** 1.625 m
37. a. 3π cm   **b.** 9.42 cm   **39. a.** 21π ft   **b.** 65.94 ft   **41. a.** 16π mm   **b.** 50.24 mm   **43.** 4 ft^2 **45.** 6 in.^{2} **47.** 15 m^2
49. $\frac{5}{8}$ ft^2 or 90 in.^{2} **51. a.** 9π mm^2 **b.** 28.26 mm^2 **53. a.** 6.25π m^2 **b.** 19.625 m^2 **55.** 84 in.^{2} **57.** 320 ft^2 **59.** 45 in.^{2}
61. 366 ft^2 **63.** 8 in.^{3} **65.** 8 in.^{3} **67. a.** 192 in.^{3} **b.** 208 in.^{2} **69. a.** 30 ft^3 or 51,840 in.^{3} **b.** 62 ft^2 or 8928 in.^{2}
71. a. 28π yd^3, 87.92 yd^3 **b.** 36π yd^2, 113.04 yd^2

CHAPTER 7 PRACTICE

1. True   **3.** True   **5.** True   **7.** True   **9.** True   **11.** True   **13.** $180°$ **15.** $540°$ **17.** $1080°$ **19.** 4 units
21. $P = 18$ ft, $A = 18$ ft^2 **23.** $w = 30$ in., $A = 480$ in.^{2} **25.** $l = 8$ cm, $w = 8$ cm, $P = 32$ cm   **27.** 4 ft^2
29. $d = 10$ in., $C = 10\pi$ in., $A = 25\pi$ in.^{2} **31.** $r = 18$ m, $d = 36$ m, $A = 324\pi$ m^2 **33.** $r = 4$ cm, $d = 8$ cm, $A = 50.24$ cm^2
35. 4 to 5 or 4:5 or $\frac{4}{5}$ **37.** $21\frac{1}{3}$ m^2 **39.** 9 cm   **41. a.** $2\pi\sqrt{2}$ cm   **b.** 2π cm^2 **c.** $(8 - 2\pi)$ cm^2 **43.** $w = 2$ ft, $h = 2$ ft, $V = 8$ ft^3
45. $V = 24$ m^3 **47.** $d = 5$ ft, $V = \dfrac{125\pi}{6}$ ft^3 **49.** $r = 3$ m, $d = 6$ m   **51.** $r = 4$ in., $h = 7$ in.   **53.** 27 times   **55.** 144π sq. units
57. 16π ft^3 **59.** $(16 + 2\pi)$ m   **61.** 4π in.^{2} **63.** $30°$

APPENDIX A

TABLES

TABLE 1 Squares and Square Roots

Number N	Square N^2	Square Root $\sqrt{N}$	Number N	Square N^2	Square Root $\sqrt{N}$	Number N	Square N^2	Square Root $\sqrt{N}$
0	0	0	35	1225	5.916	70	4900	8.367
1	1	1	36	1296	6	71	5041	8.426
2	4	1.414	37	1369	6.083	72	5184	8.485
3	9	1.732	38	1444	6.164	73	5329	8.544
4	16	2	39	1521	6.245	74	5476	8.602
5	25	2.236	40	1600	6.325	75	5625	8.660
6	36	2.449	41	1681	6.403	76	5776	8.718
7	49	2.646	42	1764	6.481	77	5929	8.775
8	64	2.828	43	1849	6.557	78	6084	8.832
9	81	3	44	1936	6.633	79	6241	8.888
10	100	3.162	45	2025	6.708	80	6400	8.944
11	121	3.317	46	2116	6.782	81	6561	9
12	144	3.464	47	2209	6.856	82	6724	9.055
13	169	3.606	48	2304	6.928	83	6889	9.110
14	196	3.742	49	2401	7	84	7056	9.165
15	225	3.873	50	2500	7.071	85	7225	9.220
16	256	4	51	2601	7.141	86	7396	9.274
17	289	4.123	52	2704	7.211	87	7569	9.327
18	324	4.243	53	2809	7.280	88	7744	9.381
19	361	4.359	54	2916	7.348	89	7921	9.434
20	400	4.472	55	3025	7.416	90	8100	9.487
21	441	4.583	56	3136	7.483	91	8281	9.539
22	484	4.690	57	3249	7.550	92	8464	9.592
23	529	4.796	58	3364	7.616	93	8649	9.644
24	576	4.899	59	3481	7.681	94	8836	9.695
25	625	5	60	3600	7.746	95	9025	9.747
26	676	5.099	61	3721	7.810	96	9216	9.798
27	729	5.196	62	3844	7.874	97	9409	9.849
28	784	5.292	63	3969	7.937	98	9604	9.899
29	841	5.385	64	4096	8	99	9801	9.950
30	900	5.477	65	4225	8.062	100	10,000	10
31	961	5.568	66	4356	8.124			
32	1024	5.657	67	4489	8.185			
33	1089	5.745	68	4624	8.246			
34	1156	5.831	69	4761	8.307			

TABLE 2

U.S. Measures	Metric Measures

Length

1 ft = 12 in.	in.: inch(es)	
1 yd = 3 ft	ft: foot (feet)	
1 mi = 1760 yd	yd: yard(s)	
1 mi = 5280 ft	mi: mile(s)	

Length

1 cm = 10 mm	mm: millimeter(s)
1 m = 100 cm	cm: centimeter(s)
1 km = 1000 m	m: meter(s)
	km: kilometer(s)

Capacity

1 tsp = 80 gtt	gtt: drop(s)
1 tbsp = 3 tsp	tsp: teaspoon(s)
1 fl oz = 2 tbsp	tbsp: tablespoon(s)
1 c = 8 fl oz	fl oz: fluid ounce(s)
1 pt = 2 c	c: cup(s)
1 qt = 2pt	pt: pint(s)
1 gal = 4 qt	qt: quart(s)
	gal: gallon(s)

Capacity

1 L = 1000 mL	mL: milliliter(s)
1 kL = 1000 L	L: liter(s)
	kL: kiloliter(s)

Weight

1 lb = 16 oz	oz: ounce(s)
1 T = 2000 lb	lb: pound(s)
	T: ton(s) or short ton(s)

Mass (Weight)

1 g = 1000 mg	mg: milligram(s)
1 kg = 1000 g	g: gram(s)
1 t = 1000 kg	kg: kilogram(s)
	t: tonne(s) or metric ton(s)

Area

$1 \text{ ft}^2 = 144 \text{ in.}^2$	in.^2: square inch(es)
$1 \text{ yd}^2 = 9 \text{ ft}^2$	ft^2: square foot (feet)
$1 \text{ A} = 4840 \text{ yd}^2$	yd^2: square yard(s)
$1 \text{ mi}^2 = 640 \text{ A}$	A: acre(s)
	mi^2: square mile(s)

Area

$1 \text{ cm}^2 = 100 \text{ mm}^2$	mm^2: square millimeter(s)
$1 \text{ m}^2 = 10{,}000 \text{ cm}^2$	cm^2: square centimeter(s)
$1 \text{ ha} = 10{,}000 \text{ m}^2$	m^2: square meter(s)
$1 \text{ km}^2 = 100 \text{ ha}$	ha: hectare(s)
	km^2: square kilometer(s)

Volume

$1 \text{ ft}^3 = 1728 \text{ in.}^3$	in.^3: cubic inch(es)
$1 \text{ yd}^3 = 27 \text{ ft}^3$	ft^3: cubic foot (feet)
	yd^3: cubic yard(s)

Volume

$1 \text{ cm}^3 = 1000 \text{ mm}^3$	mm^3: cubic millimeter(s)
$1 \text{ m}^3 = 1{,}000{,}000 \text{ cm}^3\text{(cc)}$	$\text{cm}^3\text{(cc)}$: cubic centimeter(s)
	m^3: cubic meter(s)

Temperature

Water boils at 212°F.   °F: degrees Fahrenheit
The normal human body temperature is 98.6°F.
Water freezes at 32°F.

Temperature

Water boils at 100°C.   °C: degrees Celsius
The normal human body temperature is 37°C.
Water freezes at 0°C.

Time

1 min = 60 sec	sec: second(s)	1 hr = 60 min	min: minute(s)	1 business month = 30 days	
1 da = 24 hr	hr: hour(s)	1 wk = 7 da	wk: week(s)	1 business year = 360 days	
1 yr = 12 mo	mo: month(s)	1 yr ≈ 365 da	da: day(s)		
			yr: year(s)		

TABLE 3 Conversion Factors (U.S./Metric)

	From	To	Multiply By
Length	inches (in.)	millimeters (mm)	**25.4***
	inches	centimeters (cm)	**2.54**
	feet (ft)	meters (m)	**0.3048**
	yards (yd)	meters	**0.9144**
	miles (mi)	kilometers (km)	1.609
Capacity	drops (gtt)	milliliters (mL)	16.2
	teaspoons (tsp)	milliliters	4.93
	tablespoons (tbsp)	milliliters	14.8
	fluid ounces (fl oz)	milliliters	29.6
	cups (c)	liters (L)	0.237
	pints (pt)	liters	0.473
	quarts (qt)	liters	0.946
	gallons (gal)	liters	3.79
Weight (Mass)	ounces (oz)	grams (g)	28.4
	pounds (lb)	kilograms (kg)	0.454
	tons (T)	tonnes (t)	0.907
Area	square inches (in.^{2})	square centimeters (cm^2)	6.45
	square feet (ft^2)	square meters (m^2)	0.0929
	square yards (yd^2)	square meters	0.836
	square miles (mi^2)	square kilometers (km^2)	2.59
	acres (A)	hectares (ha)	0.405
Volume	cubic inches (in.^{3})	cubic centimeters (cm^3 or cc)	16.4
	cubic feet (ft^3)	cubic meters (m^3)	0.0283
	cubic yards (yd^3)	cubic meters	0.765
Temperature	degrees Fahrenheit (°F)	degrees Celsius (°C)	0.556 (after subtracting 32)

* All conversion factors in bold type are exact. All others are rounded.

TABLE 4 Conversion Factors (Metric/U.S.)

	From	To	Multiply By
Length	millimeters (mm)	inches (in.)	0.03937
	centimeters (cm)	inches	0.3937
	meters (m)	feet (ft)	3.280
	meters	yards (yd)	1.094
	kilometers (km)	miles (mi)	0.6214
Capacity	milliliters (mL)	drops (gtt)	0.0616
	milliliters	teaspoons (tsp)	0.203
	milliliters	tablespoons (tbsp)	0.0676
	milliliters	fluid ounces (fl oz)	0.0338
	liters (L)	cups (c)	4.23
	liters	pints (pt)	2.11
	liters	quarts (qt)	1.06
	liters	gallons (gal)	0.264
Mass (Weight)	grams (g)	ounces (oz)	0.0353
	kilograms (kg)	pounds (lb)	2.21
	tonnes (t)	tons (T)	1.10
Area	square centimeters (cm^2)	square inches (in.^{2})	0.155
	square meters (m^2)	square feet (ft^2)	10.8
	square meters	square yards (yd^2)	1.20
	square kilometers (km^2)	square miles (mi^2)	0.386
	hectares (ha)	acres (A)	2.47
Volume	cubic centimeters (cm^3)	cubic inches (in.^{3})	0.0610
	cubic meters (m^3)	cubic feet (ft^3)	35.3
	cubic meters	cubic yards (yd^3)	1.31
Temperature	degrees Celsius ($^\circ$C)	degrees Fahrenheit ($^\circ$F)	**1.8*** (then add 32)

* All conversion factors in bold type are exact. All others are rounded.

TABLE 5 Geometry Formulas

Figure	Perimeter (P)	Area (A)
Square	$P = 4s$	$A = s^2$
Rectangle	$P = 2(l + w)$	$A = lw$
Parallelogram	$P = 2(a + b)$	$A = bh$
Triangle	$P = a + b + c$	$A = \dfrac{1}{2}bh$

	Circumference (C)	Area (A)
Circle	$C = \pi d$ $C = 2\pi r$	$A = \pi r^2$

	Volume (V)	Surface Area (SA)
Cube	$V = e^3$	$SA = 6e^2$
Rectangular Box	$V = lwh$	$SA = 2(lw + lh + wh)$
Cylinder	$V = \pi r^2 h$	$SA = 2\pi r(r + h)$
Sphere	$V = \dfrac{4}{3}\pi r^3$	$SA = 4\pi r^2$

TABLE 6 Primes and Prime Factorizations (2–200)

N	Prime Factorization	N	Prime Factorization	N	Prime Factorization	N	Prime Factorization	N	Prime Factorization
		41	p	81	3^4	121	11^2	161	$7 \cdot 23$
2	$p*$	42	$2 \cdot 3 \cdot 7$	82	$2 \cdot 41$	122	$2 \cdot 61$	162	$2 \cdot 3^4$
3	p	43	p	83	p	123	$3 \cdot 41$	163	p
4	2^2	44	$2^2 \cdot 11$	84	$2^2 \cdot 3 \cdot 7$	124	$2^2 \cdot 31$	164	$2^2 \cdot 41$
5	p	45	$3^2 \cdot 5$	85	$5 \cdot 17$	125	5^3	165	$3 \cdot 5 \cdot 11$
6	$2 \cdot 3$	46	$2 \cdot 23$	86	$2 \cdot 43$	126	$2 \cdot 3^2 \cdot 7$	166	$2 \cdot 83$
7	p	47	p	87	$3 \cdot 29$	127	p	167	p
8	2^3	48	$2^4 \cdot 3$	88	$2^3 \cdot 11$	128	2^7	168	$2^3 \cdot 3 \cdot 7$
9	3^2	49	7^2	89	p	129	$3 \cdot 43$	169	13^2
10	$2 \cdot 5$	50	$2 \cdot 5^2$	90	$2 \cdot 3^2 \cdot 5$	130	$2 \cdot 5 \cdot 13$	170	$2 \cdot 5 \cdot 17$
11	p	51	$3 \cdot 17$	91	$7 \cdot 13$	131	p	171	$3^2 \cdot 19$
12	$2^2 \cdot 3$	52	$2^2 \cdot 13$	92	$2^2 \cdot 23$	132	$2^2 \cdot 3 \cdot 11$	172	$2^2 \cdot 43$
13	p	53	p	93	$3 \cdot 31$	133	$7 \cdot 19$	173	p
14	$2 \cdot 7$	54	$2 \cdot 3^3$	94	$2 \cdot 47$	134	$2 \cdot 67$	174	$2 \cdot 3 \cdot 29$
15	$3 \cdot 5$	55	$5 \cdot 11$	95	$5 \cdot 19$	135	$3^3 \cdot 5$	175	$5^2 \cdot 7$
16	2^4	56	$2^3 \cdot 7$	96	$2^5 \cdot 3$	136	$2^3 \cdot 17$	176	$2^4 \cdot 11$
17	p	57	$3 \cdot 19$	97	p	137	p	177	$3 \cdot 59$
18	$2 \cdot 3^2$	58	$2 \cdot 29$	98	$2 \cdot 7^2$	138	$2 \cdot 3 \cdot 23$	178	$2 \cdot 89$
19	p	59	p	99	$3^2 \cdot 11$	139	p	179	p
20	$2^2 \cdot 5$	60	$2^2 \cdot 3 \cdot 5$	100	$2^2 \cdot 5^2$	140	$2^2 \cdot 5 \cdot 7$	180	$2^2 \cdot 3^2 \cdot 5$
21	$3 \cdot 7$	61	p	101	p	141	$3 \cdot 47$	181	p
22	$2 \cdot 11$	62	$2 \cdot 31$	102	$2 \cdot 3 \cdot 17$	142	$2 \cdot 71$	182	$2 \cdot 7 \cdot 13$
23	p	63	$3^2 \cdot 7$	103	p	143	$11 \cdot 13$	183	$3 \cdot 61$
24	$2^3 \cdot 3$	64	2^6	104	$2^3 \cdot 13$	144	$2^4 \cdot 3^2$	184	$2^3 \cdot 23$
25	5^2	65	$5 \cdot 13$	105	$3 \cdot 5 \cdot 7$	145	$5 \cdot 29$	185	$5 \cdot 37$
26	$2 \cdot 13$	66	$2 \cdot 3 \cdot 11$	106	$2 \cdot 53$	146	$2 \cdot 73$	186	$2 \cdot 3 \cdot 31$
27	3^3	67	p	107	p	147	$3 \cdot 7^2$	187	$11 \cdot 17$
28	$2^2 \cdot 7$	68	$2^2 \cdot 17$	108	$2^2 \cdot 3^3$	148	$2^2 \cdot 37$	188	$2^2 \cdot 47$
29	p	69	$3 \cdot 23$	109	p	149	p	189	$3^3 \cdot 7$
30	$2 \cdot 3 \cdot 5$	70	$2 \cdot 5 \cdot 7$	110	$2 \cdot 5 \cdot 11$	150	$2 \cdot 3 \cdot 5^2$	190	$2 \cdot 5 \cdot 19$
31	p	71	p	111	$3 \cdot 37$	151	p	191	p
32	2^5	72	$2^3 \cdot 3^2$	112	$2^4 \cdot 7$	152	$2^3 \cdot 19$	192	$2^6 \cdot 3$
33	$3 \cdot 11$	73	p	113	p	153	$3^2 \cdot 17$	193	p
34	$2 \cdot 17$	74	$2 \cdot 37$	114	$2 \cdot 3 \cdot 19$	154	$2 \cdot 7 \cdot 11$	194	$2 \cdot 97$
35	$5 \cdot 7$	75	$3 \cdot 5^2$	115	$5 \cdot 23$	155	$5 \cdot 31$	195	$3 \cdot 5 \cdot 13$
36	$2^2 \cdot 3^2$	76	$2^2 \cdot 19$	116	$2^2 \cdot 29$	156	$2^2 \cdot 3 \cdot 13$	196	$2^2 \cdot 7^2$
37	p	77	$7 \cdot 11$	117	$3^2 \cdot 13$	157	p	197	p
38	$2 \cdot 19$	78	$2 \cdot 3 \cdot 13$	118	$2 \cdot 59$	158	$2 \cdot 79$	198	$2 \cdot 3^2 \cdot 11$
39	$3 \cdot 13$	79	p	119	$7 \cdot 17$	159	$3 \cdot 53$	199	p
40	$2^3 \cdot 5$	80	$2^4 \cdot 5$	120	$2^3 \cdot 3 \cdot 5$	160	$2^5 \cdot 5$	200	$2^3 \cdot 5^2$

* p means prime.

TABLE 7 Real Numbers and Properties

Numbers

Natural Numbers or *Counting Numbers* $1, 2, 3, \cdots$

Whole Numbers $0, 1, 2, 3, \cdots$

Integers $\cdots, -3, -2, -1, 0, 1, 2, 3, \cdots$

Rational Numbers All numbers that can be written as $\dfrac{a}{b}$ where a and b are integers (and $b \neq 0$).

Irrational Numbers All numbers that equal decimals which neither terminate nor repeat.

Real Numbers All rational and irrational numbers.

Properties

For all real numbers a, b, and c:

Commutative Properties	$a + b = b + a$	$ab = ba$
Associative Properties	$(a + b) + c = a + (b + c)$	$(ab)c = a(bc)$
Identity Properties	$a + 0 = 0 + a = a$	$a(1) = 1(a) = a$
Inverse Properties	$a + (-a) = -a + a = 0$	$a \cdot \dfrac{1}{a} = \dfrac{1}{a} \cdot a = 1 \ (a \neq 0)$
Distributive Property	$a(b + c) = ab + ac$ or $(b + c)a = ba + ca$	
Zero-Product Property	$ab = 0$ means $a = 0$ or $b = 0$	
Zero-Factor Property	$a(0) = 0(a) = 0$	
Subtraction Properties	$a - b = a + (-b)$	$a - 0 = a + 0 = a$
	$0 - a = 0 + (-a) = -a$	$a - b = 0$ means $a = b$
Division Properties	$\dfrac{a}{b} = \dfrac{1}{b} \cdot a = a \cdot \dfrac{1}{b} = b\overline{)a}$	$0 \div a = \dfrac{0}{a} = 0 \ (a \neq 0)$
	$a \div 0$ or $\dfrac{a}{0}$ is not defined	$a \div 1 = \dfrac{a}{1} = a$
	$\dfrac{a}{b} = 1$ means $a = b \ (a \neq 0$ and $b \neq 0)$	
Negative Properties	$-(-a) = a$	$-1(a) = a(-1) = -a$
	$a(-b) = (-a)b = -(ab) = -ab$	$(-a)(-b) = ab$
	$\dfrac{-a}{b} = \dfrac{a}{-b} = -\dfrac{a}{b} = -\dfrac{-a}{-b}$	$\dfrac{a}{b} = \dfrac{-a}{-b} = -\dfrac{-a}{b} = -\dfrac{a}{-b}$

INDEX